Praise for *The Business of Engineering*

"Lots of great career development advice for engineering students, particularly for those entering industry."

—Jennifer Sinclair Curtis,
Distinguished Professor of Chemical Engineering,
Dean of the College of Engineering, University of California, Davis

"Matthew Loos sounds a clarion call to engineering students and engineering school faculty alike. Engineering curricula need to evolve and improve to maximize the competitiveness of engineering graduates in a rapidly changing economy, and Loos provides excellent examples of how the study of engineering can be broadened to produce more impactful engineering professionals. Well organized, good-natured, and readable, *The Business of Engineering* offers students the how and why of the soft-skills development they too infrequently have the opportunity to pursue in school. Employers will be grateful, but not as grateful as the engineers who take his advice."

—James E. Moore II,
Professor, Director, Transportation Engineering Program,
USC Viterbi School of Engineering, USC Price School of Public Policy,
University of Southern California

"This book demonstrates the principle that our being good at the business of engineering is based not only on the understanding of numbers but also on problem solutions with an understanding, ethical treatment of people. It makes it all worthwhile."

—C. Larry Weir, PE (Arkansas)
Cofounder, Hawkins-Weir Engineers Inc. (Retired),
Adjunct Professor, Civil Engineering, University of Arkansas (Retired)

THE BUSINESS OF ENGINEERING

A NEW MINDSET FOR THE ENGINEER OF THE FUTURE

Matthew K. Loos, PE

The Business of Engineering: A New Mindset for the Engineer of the Future

Published by the Engineering Management Institute

ISBN: 978-0-9989987-8-7

Gear backgrounds designed by starline / Freepik
Chapter summary background designed by davidzydd / Freepik

Book Design by James Woosley, FreeAgentPress.com

DEDICATION

To my parents, who instilled the values of hard work and perseverance within me at an early age.

To my wife, Lauren, who gave me the faith and confidence to know in my heart that I can accomplish anything in life.

To my dog, Beau, who kept my feet warm while I wrote this book every morning.

CONTENT

FOREWORD

I THOROUGHLY ENJOYED MY CAREER as an engineer; however, one day it hit me: I wanted to transition from doing the engineering myself to helping engineers become more successful. How could I do that? Where did engineers struggle the most? I found the answer to that question to be in the transition from engineer to manager.

Most engineers get promoted to manager because they are good at engineering, not because they are good at managing.

So, back in 2009, I took a leap of faith. I left my comfortable engineering job, started the Engineering Management Institute (EMI), and have traveled the world since, helping engineers build the skills needed to become effective managers and powerful leaders.

Along the way, thanks in part to the content we have created, including multiple podcasts that we publish, I have had the good fortune of meeting and speaking with motivated engineers around the globe.

That is how I met Matthew Loos. One day, Matt sent me a message through LinkedIn thanking me for the work we are doing at EMI and asking for advice on publishing a book he had written entitled *The Business of Engineering*.

Now, we have thousands of readers and podcast listeners, but I had never received a message like this from one of them. So I asked Matt if he would send me the first chapter of the book so I could read through it and provide some advice. He did, and after I read it, I quickly asked for the next chapter and eventually read through the entire manuscript in less than a week.

Immediately after finishing it, I asked Matt if he would be open to EMI publishing the book, since I felt it really captured everything I thought engineers need to be successful in today's world—working on complex projects and trying to keep up with fast-changing technology.

He said yes, and here it is. In a world of information overload, *The Business of Engineering* provides a practical framework that engineers can apply in their careers to dramatically increase their success.

I am thrilled that Matthew Loos took the time to write this book, thrilled that we at EMI were able to publish it, and most excited that you have decided to incorporate it into your professional development efforts.

Thank you to all of those people, including our Kickstarter backers, who helped turn Matthew's experience and research into a tool to be used by others.

Enjoy the book.

I have no doubt it will help you engineer your own success!

To your success,

Anthony Fasano, PE
President & CEO of the Engineering Management Institute
Author of *Engineering Your Own Success*

“The path to the CEO’s office should not be through the CFO’s office, and it should not be through the marketing department. It needs to be through engineering and design.”

—ELON MUSK, FOUNDER OF SPACEX

INTRODUCTION

ARE YOU AN ENGINEERING student sitting in the last class of your college career, wondering when you will ever use all of those engineering classes? Or are you a few years into your budding engineering career and feeling stuck in a rut, clueless about how to get ahead in your chosen field of study?

In this book, I will discuss the importance of graduating with that coveted engineering degree. I will also strive to help you understand how engineers can use various sought-after business techniques to accelerate their careers. Lastly, I will provide ways in which you can unleash your full potential by incorporating the skills of two seemingly distinct career fields: business and engineering.

According to a study done by SpencerStuart, "33 percent of the S&P 500 CEOs' undergraduate degrees are in engineering and only 11 percent are in business administration." Some of these CEOs include Amazon's Jeff Bezos, SpaceX's Elon Musk, Henry Ford, and General Motors's Mary Barra. While studying for numerous business classes in school, I became preoccupied with these visionary business leaders. What separated these men and women from the rest of the engineers throughout history? What made them so successful in the businesses they formed and managed? How could I become as successful (or more so) as these incredible engineers? Needless to say, I came away with more questions than answers. As famous English writer Sydney Smith once wrote, "What you don't know would make a great book." This book was my quest for a clearer vision of success in engineering. This book was written to answer these questions for myself and hopefully for others such as yourself. It's a "battle cry," in a sense, for engineers who are interested in becoming successful not only in engineering but also in business.

This book is divided into three parts. Section 1 will focus primarily on the story behind the book. The Business of Engineering will be dissected by looking at real-life case studies. We will begin by learning about engineers who paved the way for business success. Engineers like Tesla, Ford, and Musk help provide clues into what it takes to be successful in not only engineering but also in business. We will look at their lives in brief and then discuss what attributes were most critical to their success. As you will see from the stories presented, these men and women had no more experience than most engineers in business. After understanding the background of this concept, the reasons why the Business of Engineering is such an important mentality to have in this fourth industrial revolution will be clear. What is it about this era that's so exciting when it comes to engineering and business? How is the rate at which technology is advancing so important for those in engineering? After plunging into the importance of these skills, a discussion will follow that gives you, the reader, an introduction in how to set yourself up for success in engineering and business by following the information presented in the next two sections.

Throughout sections two and three, the engineering abilities I learned in formal engineering school, along with the business strategies learned in the real world as a consulting engineer, business owner, and real estate investor, will be discussed. I often hear the same complaint about college: "I never use anything I've learned in college for my actual job." Although this may be partially true, the knowledge and contacts gained while in college are extremely important in the real world. Even though many of the classes taken by engineers and business majors are very much different, the analytical thinking skills are very much the same. Engineers and businesspeople have unique skill sets that can overlap and produce powerful results in your career. My goal is to highlight those skill sets for the reader. A great mentor of mine once told me that we were not just engineers, "we were in the business of engineering." My overarching goal in this book is to give you, the reader, a deeper meaning of that statement.

According to a recent study, only 60 percent of the books purchased ever get read. Even more startling is that once opened, some books only get read to completion about 20 to 40 percent of the time.

My hope for you is that you follow along on this journey to completion. I know there may be outside distractions that tempt you to set this book down, but if you're the type who bought this book, I imagine you are the type of person who doesn't give up easily. You are most likely the type who strays away from the crowd. I commend you for wanting to further your professional skills. My hope for you is that you can use my advice and the counsel of my many mentors going forward to help realize your full potential in engineering and business ventures.

As we end the introduction of this book, another chapter begins. You have two options at this point: close the book and put it back on the shelf or commit to advancing your career by reading on.

> ***"There are two mistakes one can make along the road to truth—not going all the way, and not starting."***
>
> **—BUDDHA**

"You can't connect the dots looking forward; you can only connect them looking backward. So you have to trust that the dots will somehow connect in your future. You have to trust in something—your gut, destiny, life, karma, whatever. This approach has never let me down, and it has made all the difference in my life."

—STEVE JOBS

SECTION 1:

ENGINEERING IN THE FOURTH INDUSTRIAL REVOLUTION

"The object of life is not to be on the side of the majority, but to escape finding oneself in the ranks of the insane."

—MARCUS AURELIUS

CHAPTER 1:

WHAT IS THE BUSINESS OF ENGINEERING?

IF YOU'RE READING THIS book right now, chances are you purchased it on Amazon or through one of its affiliates. With net sales of $107 billion and 230,800 employees at the time of this writing, Amazon has now overtaken the top spot on the charts as the largest online retailer. The engineer-turned-entrepreneur who made this possible is none other than Jeff Bezos. Bezos developed the idea for this e-commerce site in 1995. Little did he truly know the magnitude of this venture when he began.

Bezos started out with a strong interest in tearing things apart and putting them back together in his home "laboratory," just like any other normal burgeoning engineer. He was born in 1964 in New Mexico and, as a teenager, moved to Miami with his family. His interest in computers and electronics continued to grow. He also began to enjoy making money. According to Biography.com, "It was during high school that he started his first business, the Dream Institute, an educational summer camp for fourth, fifth, and sixth graders." This venture would be just the beginning of his successful business career.

"What we need to do is always lean into the future; when the world changes around you and when it changes against you—what used to be a tailwind is now a headwind—you have to lean into that and figure out what to do because complaining isn't a strategy."

—JEFF BEZOS

After graduating as the valedictorian of his high school, he attended Princeton University. With his growing affinity for technology, Bezos decided to pursue dual degrees in computer science and electrical engineering. Not a bad set of degrees to pursue in the late '80s. He would go on to graduate summa cum laude in 1986. After graduating with such a prestigious set of degrees, he went to work for D. E. Shaw & Co., a Wall Street investment firm. Four years later, he became senior vice president. You read that right. Only four years later he earned the title of senior vice president. Some would say he'd made it to the big leagues or that he'd achieved the American dream; however, he still yearned for something different. Deep down, he had an entrepreneurial spirit that needed to be unleashed.

He quit the firm and made the move to Seattle to create his start-up. Seattle at the time was going through a great renaissance. The up-and-coming Seattle-based rock band Nirvana was starting to sweep the nation. Many software engineers were racing to work there and be a part of the growing tech industry already present in this once sleepy West Coast town. Things were good in Seattle. There was also another reason that Seattle was a prime candidate for where to start his fledgling business. Unlike California, which has over 8 percent corporate income tax and over 13 percent personal income tax, Washington has neither. That's right, Washington boasts neither individual nor corporate income taxation. Washington therefore is a great state in which to start a new business. Bezos knew this. All of the factors, including a blossoming tech scene, a growing cultural identity, and nonexistent income taxation helped to make the decision to move to Seattle an easier one for Bezos.

As most of you may know, Amazon started out by selling just books, and it soon became the largest book retailer in the world. This should come as no surprise, seeing that he could sell millions of titles online, whereas a brick-and-mortar shop could only sell thousands, if it was big enough. Now you know that Amazon sells much more than books and has become the largest online retailer in the world. This is not the path you would imagine for an electrical engineer, but he had the vision to create and capitalize on the Internet craze at the time.

Jeff Bezos's story epitomizes the central theme of this book. Bezos could never have reached the heights he has experienced with Amazon without a background in business. To become a senior vice president of an investment firm in four years takes much more than technical skills alone. This is where the Business of Engineering story emerges. The Business of Engineering is more about the trajectory in which the field of engineering is trending. As the global economy becomes even more intertwined, the technological skill set provided by engineers will be even more necessary in day-to-day business management decisions. We are seeing this trend come full circle from its time in the industrial age with engineers-turned-business moguls such as Henry Ford, Thomas Edison, and Nikola Tesla. Once the industrial age was behind us, the engineer was pushed back into the role of research and development, and the trained businesspeople reclaimed their positions in business leadership roles.

Only recently have we seen a reemergence of engineers in the upper-management positions they vacated previously in history. There has been a surge in technological advances that have put engineers back in these leadership positions. Some examples of these present-day engineering leaders are Elon Musk, Jeff Bezos, and Mary Barra.

What do all of these engineers-turned-business-moguls have in common? What has separated these engineers from the rest? To answer these questions, we need to look at the intangibles. They all possessed an unparalleled technical proficiency, but there was something much more. They all utilized skills not easily taught in schools. These abstract skills are often referred to as "soft skills." Soft skills are defined by Oxford Dictionaries as "personal attributes that enable someone to interact effectively and harmoniously with other people."

Similar to the way software coordinates the various pieces of hardware in a computer, soft skills help to coordinate your various "hard skills" with other people. Soft skills are the nontechnical bridge in the gap between one's technical knowledge and the understanding of others. Now that we have a broad idea of what soft skills are, let's dig into the actual skills that are often considered soft skills. For that, I'd like to use some of the most renowned engineers-turned-business moguls as examples to fuel the discussion. The list of engineers who fit the bill includes Thomas Edison, Henry Ford, Nikola Tesla, Jeff Bezos, Elon Musk, and Mary Barra.

The most common trait that all of these incredible engineers possess is an exceptional ability to solve problems. The ability to solve complex societal problems helped them to capitalize on their skill sets. These business moguls also utilized big-picture thinking to look at problems from a different perspective than most people. They each possessed technical skills that gave them a unique perspective and understanding of the market they controlled. Each one of these candidates viewed potential problems through both an analytical and logical lens that was often underutilized in the business world. Throughout history, these engineers used a detail-oriented approach to tackle some of the problems facing their future customers. Each one also possessed a strong work ethic. They understood that hard work and persistence would pay dividends. Finally, these engineers displayed a natural talent for leadership roles.

In the subsequent pages, we will discuss each one of these qualities and use real-life case studies to show how these traits catapulted each of these leaders to the top. We will pull back the curtain on the lives of these engineering-educated rock stars to see what really attributed to their individual successes.

Problem-Solving

From Thomas Edison to Mary Barra, the ability to solve problems was a common trait in engineers-turned-business moguls. Their problem-solving techniques helped them discover the benefits of the alternating current, how to utilize manufacturing techniques, and even how to exploit the Internet to make literally any commodity available to anyone willing to purchase it.

Each of these engineers possessed a unique curiosity to understand problems that most people didn't even know they had. Henry Ford stated, "If I had asked people what they wanted, they would have said faster horses." Looking back on this statement, we can see the inherent hilarity of it, but it was an honest statement. This quote shows how Henry Ford was able to solve problems that most people didn't even realize existed. When most people thought that horses were the pinnacle of transportation, Ford proved them wrong. In a time when people thought that horses could not be topped in human locomotion for the common man, Ford dared to believe that the automobile would move the masses.

Jeff Bezos understood that the Internet was a powerful tool that could be harnessed to spread knowledge throughout the world. He could produce the world's largest bookstore to help spread ideas and knowledge around the world at a time when books weren't as accessible to those who lived in more remote locations. The value he added to the market has been shown through the growth of the world's "everything" store. Mary Barra has helped bring General Motors—one of the oldest automotive producers—into the twenty-first century. Since becoming CEO, Mary has acquired Cruise Automation, launched GM's first production electric vehicle, and invested heavily in the ride-hailing service Lyft. Needless to say, she was out to solve the problems that most of the previous GM CEOs did not see coming. Her forward-thinking attitude has propelled her to the top. She is willing to take on problems and challenges that most of the previous leaders at GM would not even touch.

Problem-solving skills were the catalyst in these leaders' rise to the top. When you begin to solve problems for the masses—better described as adding value to the marketplace—the market will reward you. These engineers added value to the marketplace by solving problems for the masses and were thus rewarded. Problem-solving skills weren't the only way these engineers became successful; big-picture thinking helped them to see what others could not.

Big-Picture Thinking

Problem-solving is best done at a higher level. As stated previously, these engineers were able to anticipate problems that most of the general public didn't even know existed. They were able to look at society from a thirty-thousand-foot view and solve problems that were often hidden in plain sight. As I mentioned before, most people didn't imagine a better means of transportation than the horse until Henry Ford proposed it. Elon Musk has enamored many sci-fi fans with his push for human exploration of Mars. He truly believes life on other planets is the key to the human race's longevity. This is real big-picture thinking. Mary Barra is constantly pushing the envelope in her role at GM. She understands that the automobile industry as a whole will not be sustainable unless it takes measures to reduce fossil fuel consumption and dependency. She has led the march in producing a mass-manufactured electric vehicle to compete with Elon Musk's Tesla.

Big-picture thinking is one of the most undervalued skill sets that engineers can bring to the boardroom. Managing and growing a company requires vision. The CEO's main goal is to align the company in such a way that the growth of the company aligns with its long-term projections. One way to do this is by harnessing and wielding big-picture thinking.

> ***"When Henry Ford made cheap, reliable cars, people said, 'Nah, what's wrong with a horse?' That was a huge bet he made, and it worked."***
>
> ***—ELON MUSK***

Technical Proficiency

Technical proficiency is more than just knowing "how things work." It's a fundamental skill needed for a deeper understanding of certain processes. In all of the case studies mentioned so far, each engineer understood the intricacies of the technologies they were promoting. They had a deep understanding of how things worked and were able to speak competently with the people who were actively utilizing and building these technologies. In many cases, these leaders with technological backgrounds are able to understand what is truly best for their outputs. This helps to separate them from their non-technologically savvy counterparts.

This is the one truly unique skill engineers can apply without even trying. In engineering school, students use various methods of problem-solving to understand how and why things work. This is the underlying principle for obtaining technical proficiency. Technical skills are constantly drilled into engineering students. Traditionally, this skill set is more specialized into certain fields based on the engineering discipline the student chose, but the foundation for this skill is still very useful when learning other processes or technologies.

I believe this technical proficiency comes from an inherent curiosity in almost all engineers—this curiosity that was cultivated as a child playing with Legos and other mechanically stimulating toys. Technical proficiency can be learned, but engineers have a great advantage by actually enjoying the process of learning technical skills. Most non-engineers or creative types don't quite understand this enjoyment, and I can't quite explain it. What I can say is this is the main reason why I entered the field of engineering. An enjoyment of technical processes and a love of math and science are what sparked me to pursue the engineering profession. These attributes of not just technical proficiency—but technical enjoyment—make the engineer invaluable in the boardroom.

I also believe this attribute can be learned for the majority of businesspeople who just want a leg up on the competition. By getting to know your market and your product or service, you can gain the technical proficiency needed to create opportunities for yourself. When you know your product and your process, it becomes much easier to convince others to buy into it.

Detail Oriented

"Don't sweat the details" was not a phrase used often by these engineers. Being detail oriented is not a skill that comes easy. I often find myself struggling with this very skill in my day-to-day experiences. When I'm working on several projects at once, I find that it's often hard to focus on the details. I sometimes have to stop myself and refocus my mind. I can't imagine handling all of the projects that these engineers-turned-business moguls have taken on at once. These engineers understood that details could make or break their businesses and ultimately their dreams.

To do the same, you must start thinking about details a little more intensely. By prioritizing your tasks and concentrating on the task at hand, you can drastically reduce errors in your own work. In Chapter 6, we will delve much deeper into this topic.

Strong Work Ethic

These engineers were not overnight successes. It's easy to believe that such success stories happened in a brief period of time, because most autobiographies only take a few hours to read. These business moguls, no matter their age, worked tirelessly to reach the pinnacle of their careers. From the time Ford produced his first "horseless carriage" to when the first Model T rolled off the assembly line, twelve years passed. Mary Barra worked for General Motors throughout all facets of the company over thirty-four years before becoming CEO.

As you can see from these timelines, these engineers were not overnight successes. It's easy to read a biography in a few hours and think that it doesn't seem that hard to replicate what they did. That two-hundred-page book does little justice when describing the hard work and long hours that these engineers put into their craft. They spent countless years working to achieve success. Unfortunately, or fortunately for you, if you want to be massively successful, you will have to do the same.

Leadership

Leadership was thrust upon these business magnates. They were never formally taught the basics of leadership in a classroom. They learned the ins and outs of leadership through trial and error. Each one of these successful leaders had their own leadership style, but all were effective when you see how far-reaching their influence was.

These men and women utilized all of the skills I've mentioned so far, in some combination, to become successful in their various endeavors. When you scan the history books, you will see many more of the same stories from other engineers who broke through and achieved phenomenal careers. These leadership skills provided them with the self-confidence to tackle all sorts of challenges, even the ones that seemed most out of reach to the greater population. Unfortunately, most of these skills cannot be taught in a classroom setting; they have to be learned through other means.

To understand and fully utilize these attributes, you must shift your thinking. Before reading the biographies of so many successful engineers, entrepreneurs, and businesspeople, I had only a glimpse of what was necessary to become successful. The next couple of chapters will help you reboot your thoughts and form new neural pathways to help you think like the successful engineers referenced in this book. Once you let go of your preconceived notions of success, you'll open your mind to the true possibilities.

The Business of Engineering isn't just a phrase; it's a mindset that can be adopted by anyone who truly wants to take their career performance to the next level. There is a tremendous need in the marketplace for those who can marry the attributes of engineering and business in their professional lives. Why is it so important for us to adapt this mindset? What specifically are employers looking for? How can the skills learned in engineering school prepare you for success?

CHAPTER 1 SUMMARY

WHAT IS THE BUSINESS OF ENGINEERING?

The Business of Engineering is a mindset held by some of the most successful engineers and business professionals in the world. The attributes that define the Business of Engineering are composed of the commonalities of each profession—business and engineering. When meshed together, the powers of each profession combine to catapult one's career to the next level.

This is where the Business of Engineering story emerges. As the global economy becomes even more intertwined, the technological skillset provided by engineers will be even more necessary in day-to-day business management decisions. In contrast, the wildly expanding breadth of technological advances will require the modern engineer to utilize common business proficiencies to communicate these advances.

The fourth industrial revolution is an exciting time for engineers and business professionals alike. This era will be unlike any before it with the introduction of artificial intelligence (AI), supercomputers, and blockchain technologies. Society will need authorities from both the engineering and business professions to positively influence the trajectory of our society for future generations.

"I'm an engineer by trade, and what engineers do is they go and build, and they don't think a lot about storytelling."

—TRAVIS KALANICK

CHAPTER 2:

THE ENGINEER OF THE FUTURE

BEHIND THE SCENES, CORPORATIONS have been utilizing more and more nontechnical measures of potential candidate success, such as personality tests, emotional intelligence tests, cognitive ability tests, and talent assessment tests. This trend continues for employers of engineers and non-engineers alike. The need for these nontechnical soft skills has increased as employers have started to realize the importance of these skills in the workplace.

A 2015 report by the Institution of Engineering and Technology in the United Kingdom found that "the most sought-after roles in June 2015 included Engineering (No. 1) and IT and Computing (No. 4), yet hiring activity is being constrained by a lack of candidates. British Chambers of Commerce research revealed that soft skills are at the top of employers' wish lists." Another study, by NACE (National Association of Colleges and Employers) found that "more than 80 percent of responding employers said they look for evidence of leadership skills on the candidate's resume, and nearly as many seek out indications that the candidate is able to work in a team. Employers also cited written communication skills, problem-solving skills, verbal communication skills, and a strong work ethic as important candidate attributes." Another study provided by the widely used networking website LinkedIn shows that "of 291 hiring managers in the U.S. . . . 59 percent of them believe that soft skills are difficult to find." So, why is this so important to students as well as established employees?

As our capitalist society continues to flourish, you will see more engineers adopting this mentality of becoming excellent businesspeople. The current job market will force engineers to climb out of their comfort zone, because engineers must also be excellent social, as well as technical, personnel. Employers are constantly searching for talent that is both technologically savvy and socially adept to conquer the emerging markets. According to Angela Froistad, director of the College of Science and Engineering Career Center at the University of Minnesota, "Technical skills alone are not enough to ensure a successful engineering career, as engineers need to be able to function as a member of a team, think critically, and have a strong work ethic." When faced with a decision, "it's these soft skills that will differentiate candidates from one another." This should come as no surprise for most. The market is requiring more and more engineers to shake off the stereotype of being an introvert and embrace social proficiency.

As alluded to in the previous chapter, engineers are being called upon to be leaders in today's increasingly high-tech world. Engineers looking to become leaders in this ever-changing technological frontier will need to be able to not only communicate with others in technical and nontechnical settings but also solve problems that others cannot.

The so-called "engineer of the future" is described in a paper published by the National Academies of Sciences, Engineering, and Medicine (NASEM). Written in 2004, this paper summarizes the attributes that will be most necessary for engineers in 2020. A lot has changed since 2004, but this text still seems relevant for the engineer of tomorrow. The paper states that these engineers must possess **strong analytical skills, practical ingenuity, creativity, the ability to communicate effectively, dynamism, agility, resilience,** and **flexibility.** None of these skills and traits have anything to do with technical abilities such as coding and programming, technical writing, or even thermodynamics. The NASEM paper describes the engineer of the future as follows: "He or she will aspire to have the ingenuity of Lillian Gilbreth, the problem-solving capabilities of Gordon Moore, the scientific insight of Albert Einstein, the creativity of Pablo Picasso, the determination of the Wright brothers, the leadership abilities of Bill Gates, the conscience of Eleanor Roosevelt, the vision of Martin Luther King [Jr.], and the curiosity and wonder of our grandchildren."

In 2012, Google decided to investigate the reasons why certain managers were more successful than others. The company unearthed a very shocking discovery on what traits successful employees and managers possessed. Surprisingly, technical skills were last on the list of eight successful traits revealed by the multiyear study. The first seven traits of highly successful managers were:

1. The ability to be a good coach.
2. The ability to empower a team without micromanaging.
3. The ability to express interest in and concern for team members' success and personal well-being.
4. Being productive and results oriented.
5. The ability to be a good communicator who listens and shares information.
6. Being willing to help with others' career development.
7. The ability to have a clear vision and strategy for the team and the key technical skills to help advise the team.

Seven of the eight skills are what most would consider soft skills. Even in a highly technical workplace such as Google, soft skills are extremely important for employees to be successful.

The engineer of the future will need to possess skills not commonly taught in engineering schools today. However, we can see that the curricula for a majority of engineering schools are beginning to change. Many schools are starting to adopt alternative learning strategies to engage more and more bright students who may not have previously chosen engineering as a career path. The engineering schools have started to realize that technology is moving at such a rapid rate that they must prepare students differently than they have in the past. Engineers of the future must possess skills like innovation, entrepreneurial vision, and teamwork. These aren't traits that have commonly been attributed to engineering, but the rate of technological change has been so rapid that most college courses become obsolete by the time students graduate. Key intangible traits will help these graduates thrive by enabling them to learn new skills long after the college experience is over.

According to the Encyclopædia Britannica, the origin of the word "engineer" is derived as follows: "words engine and ingenious are derived from the same Latin root, *ingenerare*, which means 'to create.' The early English verb *engine* meant 'to contrive.' Thus, the engines of war were devices such as catapults, floating bridges, and assault towers; their designer was the "engine-er," or military engineer. The counterpart of the military engineer was the civil engineer, who applied essentially the same knowledge and skills to designing buildings, streets, water supplies, sewage systems, and other projects." This initial derivation of the name engineer is most definitely outdated, due in part to the massive amounts of technological change that have occurred from the industrial age to the information age. Engineers are being called upon to be more nimble and agile in the sense that the knowledge gained one day is often obsolete the next due to the rate of discovery.

As history commonly repeats itself, engineers will be shifting into more leadership roles within corporations, similar to engineers' migration into business in the industrial age. Like the industrial age, the information age is a time of immense technological innovation. In the second industrial age, engineers like Henry Ford, Nikola Tesla, and Guglielmo Marconi were all known for their skills in engineering and business. Now we're approaching our "fourth industrial revolution," as coined by Professor Klaus Schwab, founder and executive chairman of the World Economic Forum. The fourth industrial revolution, according to Schwab, "is characterized by a range of new technologies that are fusing the physical, digital, and biological worlds, impacting all disciplines, economies, and industries, and even challenging ideas about what it means to be human." This is an exciting time not only to be alive but also to be an engineer. You can see the shift that has come with this revolution with engineers such as Mary Barra, Elon Musk, and Jeff Bezos taking positions as business moguls with an ever-increasing emphasis on furthering the pursuit of technological advancement.

So why is this important? It seems fairly obvious that the engineer of the future will require many of the skills that seem counterintuitive to them. Time and technology are not standing still in this era and neither should the skill set of the modern engineer. In the next chapter, we will continue this deep dive into how engineers can create and adapt to manifest the true Business of Engineering mindset.

CHAPTER 2 SUMMARY

THE ENGINEER OF THE FUTURE

The successful engineer of the future will be one who is equipped to handle the various technological advances that will most surely come. The engineer of the future will need a solid foundation of business principles and soft skills, including:

1. Strong Analytical Skills
2. Creativity
3. Robust Communication Skills
4. Agility
5. Flexibility
6. Strong Work Ethic

On top of all this, a strong technical background will be a necessary foundation. So how do we put the principles into practice for the engineer who wants to take their career to the next level?

"A smart man makes a mistake, learns from it, and never makes that mistake again. But a wise man finds a smart man and learns from him how to avoid the mistake altogether."

—ROY H. WILLIAMS

CHAPTER 3:

MERCEDES-BENZ AND THE PATH TO SUCCESS

THE INCESSANT PATTERING OF the engine seemed to drone on for hours. As the hard rubber wheels bounced along the dirt paths, the passengers were jostled around violently. A young woman and her two teenage sons were set to create history on this drive across the rolling German countryside.

In early August 1888, Bertha Benz and her two sons were driving across the various dirt paths on the way to Pforzheim, Germany, to visit Bertha's family. Imagine seeing a large, three-wheeled horseless carriage traversing these roads once ruled by the common horse and buggy. This sight was probably akin to seeing an alien spacecraft hovering over a field.

The 2.5-horsepower motor made all sorts of noises as it struggled up the small hills on the path to Pforzheim. The sixty-five-mile automobile drive was the first of its kind and only took roughly twelve hours. This became the first long-distance automobile drive in history. Word of mouth spread far and wide about this seemingly safe and somewhat efficient horseless carriage.

The story of the production processes Henry Ford used to bring automobiles to the masses with his innovation of mass manufacturing was discussed earlier, but the story of another engineer, named Karl Benz, is much more obscure. Benz was the founder of Benz & Cie, or as we know it today, Daimler-Benz, after its merger with Gottlieb Daimler and Wilhelm Maybach.

Like many successful people throughout history, Benz had several trials and tribulations on his journey to success. He started his first company with a business partner named August Ritter when he was only twenty-seven years old. This company included an iron foundry and mechanical workshop. Soon after starting the company, his business partner became very unreliable. Benz's soon-to-be wife, Bertha, decided to put up the dowry money she was expecting to buy out August's share of the company.

During this time, Karl began experimenting with multiple new patents that would help generate some extra revenue. He formulated the design for many of the common pieces of machinery on the automobile that we now take for granted. After patenting the first two-stroke engine, he obtained patents for the speed-regulation system, spark plug, carburetor, gearshift, clutch, and water radiator. But problems would soon arise again after these successes.

The bank handling the various loans to this early business soon required an incorporation of the business due to the high production costs it was incurring. But Benz was a scrappy businessman and found two other people to join in this incorporation to satisfy the bank. The situation was not ideal, though. He only retained 5 percent of the company's ownership. Soon, he left this venture and started another company with the owners of a bike shop he frequented. The three men formed Benz & Cie. This business produced industrial machines and gas engines. Benz was able to follow his passion of creating a horseless carriage that consumed him on his long bicycle rides from years past. Two years later, he manufactured the first automobile able to produce its own power. A few months after that, Bertha Benz would take that fateful drive in history across the German countryside.

The story of Karl Benz and his successes and failures is extremely inspiring for anyone looking to make a name for themselves. His story illustrates perfectly the steps needed to implement this idea of the Business of Engineering. First and foremost, Benz changed his mindset from one of an unsatisfied employee at various industrial facilities to the mindset of a leader and a successful entrepreneur. Second, he took colossal action. He didn't take no for an answer in any of his ventures. He learned from his previous misfortunes and applied that knowledge to his next venture. Benz countered every setback with massive action. Third, he learned voraciously about everything needed to create a new life for himself and his growing family. And finally, he changed his environment. In fact, Benz did more than *change* his environment—he *created* it.

This early engineer had everything it took to manifest his passions and make them a reality. At a time when most could not fathom the idea of a carriage powered by anything but a horse, Karl Benz dreamed big. Using the action items above, Benz was able to create true success and a world that would never be the same again. The story of Karl Benz can help guide and inspire you to fully embrace the Business of Engineering in your own personal and professional endeavors.

Change Your Mindset

People thought it couldn't be done. Esteemed doctors and physicians argued about whether it was even physically possible with human skeletal structure. The sub-four-minute mile run seemed as elusive as Bigfoot or the Loch Ness Monster. Countless people had tried—and failed—to run the mile in less than four minutes. At the time, it seemed almost as improbable as walking on the moon. Then, something changed.

In 1954, a young man shattered all notions of this barrier. On May 6, 1954, Roger Bannister ran a sub-four-minute mile. Aided by his friend Chris Brasher, who acted as a pacemaker, Roger crossed the mile marker with an official time of three minutes and forty-three seconds. He believed shattering the four-minute threshold was possible. As a result, he willed his body to push harder than ever.

This phenomenon didn't last long, though. Less than two months later, an Australian named John Landy ran the mile in three minutes and fifty-eight seconds. Three years after that, an American named Don Bowden beat the four-minute mile again. Since then, over 450 Americans have broken the four-minute mile. What changed? Why had so many people suddenly started breaking the four-minute mile? Did human physical ability immediately change?

The story of Roger Bannister is one of many that shows what can happen when you don't allow the limits of what we believe is humanly possible to hold you back. The runners who all came after Bannister's record-breaking mile discovered it was truly possible to run a sub-four-minute mile. Once they saw it was conceivable, their mindset was no longer a limiting factor in their success to break this barrier.

As you begin creating your own personal and professional goals, you will need to police your thoughts so that they don't limit your possibilities. Changing your mindset will help you remove barriers you didn't originally believe you could. In the same way that Karl Benz made his wild daydreams of producing the first horseless carriage a reality, you, too, can achieve your personal and professional goals.

While in high school, I took a German class. Like almost any foreign language class, you picked a new name for yourself in the class based on the names common to that culture and region. My name was Axel. I thought this was the coolest name I could possibly pick. We were given a few days to really learn our identity. What would Axel do for fun? What kinds of foods did he like? Where did he go to school? This became a fun activity for most of the class. Some really embraced the identity they had chosen and ran with it. Others in the class thought role-playing was ridiculous and were incredulous of any possible benefits of the exercise.

By the end of the semester, it was easy to see that the ones who really fleshed out their German doppelgängers outshined their peers who were more skeptical of the exercise. When I began to see myself as a German, I began to believe I could also communicate like one. It's so easy for our minds to create alternate realities for ourselves, but too often these imagined realities are negative.

Most people like to tell themselves they are poor public speakers or that they are bad at math or that they don't like to read. Not surprisingly, most of these people never develop their public speaking skills, their proficiency in math, or even their proclivity to read. These self-limiting beliefs are holding many back from realizing their true potential. Being truly successful in your career, whether it's engineering or business, requires you to shift your perception of your abilities. Just as Karl Benz believed he could become a successful entrepreneur, or Roger Bannister believed he could tackle the four-minute mile, or even myself getting into the mindset of Axel to learn the German language, we all have the ability to change our reality by just having unwavering faith that it can be done.

Learn Voraciously

This brings me to another man who was determined to learn a new language. Arnold Schwarzenegger was born in Graz, Austria, in 1947. His father was a police officer and wanted nothing more than for his son to join the ranks of the Austrian police when he was old enough to do so, but Arnold was dreaming much bigger. By the age of thirteen, he had picked up his first dumbbell. This would change the course of history for not only a young Arnold Schwarzenegger but also the world.

> ***"For me, life is continuously being hungry. The meaning of life is not simply to exist, to survive, but to move ahead, to go up, to achieve, to conquer."***
>
> ***—ARNOLD SCHWARZENEGGER***

He began obsessively reading bodybuilding magazines and watching videos of his favorite bodybuilders, such as Reg Park and Steve Reeves. He wanted to be just like these real-life superheroes. He learned how they ate, how they worked out, and what supplements they took. His passion for learning made it possible for him to reach his goals. He competed and won his first competition in Graz in 1965. From then on, he was hooked on learning.

Schwarzenegger would take a leap of faith and move to the United States three years later and learn everything he could from the bodybuilders already training in California. The "Austrian Oak" would go on to win four Mr. Universe competitions and seven Mr. Olympia titles over the next twelve years. He had finally reached the peak of his bodybuilding career. While he was competing, he decided to pursue his childhood dream of acting, learning how to become an actor when he wasn't in the gym. He would go on to star in box office classics such as *The Terminator*, *Predator*, and *Kindergarten Cop*. After this stint in acting, he ran for governor of California in 2003. His passion for the state ultimately won over voters, and he served from 2003 to 2010. This was no doubt a learning experience for the Austria-born ambassador.

Schwarzenegger's appetite for learning new skills helped him overcome many hurdles and obstacles to achieve his goals. It's easy to become complacent in our jobs due to only having knowledge of a few specific tasks. When that happens, people stagnate. Once people leave their last class in a formal education setting, they often think they have nothing left to learn. If you would truly like to implement the Business of Engineering in your own life, you have to be willing to admit that you still have much to learn. Thinking you already know everything will only stifle your success.

Just as Karl Benz set out to learn everything he could about the modern combustion engine and the parts that would make the horseless carriage a reality, you must also embrace a passion for learning long after you walk out the classroom door for the last time.

Take Colossal Action

The word "colossal" is defined by Merriam-Webster as "of an exceptional or astonishing degree." Colossal is used to describe many things, from hamburgers at the local burger joint to the Seven Wonders of the World. Does it also describe your work effort every day?

Karl Benz made a choice to devote all his effort to creating the first horseless carriage early in his career. This choice cemented his name in the history books. He was willing to do whatever it took to make his dreams a reality. This choice brought about an understanding that he would do the work and endure any ridicule that going above and beyond often brings. To generate colossal action, you must:

1. Understand your end-game goals.
2. Understand how this action will affect your personal and professional lives; at the same time, you will have to accept what is to come, which can be uncomfortable at times.
3. Understand your "why" for taking this action; your "why" can come from a feeling of pain or pleasure.
4. Create a sense of urgency for your actions and act on them.

This list may make you a little uneasy, but unfortunately, colossal action is meant to make you feel a little uncomfortable. If taking colossal action were comfortable, everyone would do it.

One of my absolute favorite basketball players growing up was Michael Jordan. Jordan was known for his unwavering dedication to his athletic profession. I can vividly remember watching Game 5 of the 1997 NBA Finals—the famous "flu game" for Michael Jordan. Jordan was stumbling around the court, dehydrated from illness, which left him ravaged. His competitive drive compelled him to push himself. He scored thirty-eight points to help his team beat the Utah Jazz and the Chicago Bulls would go on to win the series in six games.

> ***"Some people want it to happen, some wish it would happen, others make it happen."***
>
> ***—MICHAEL JORDAN***

The actual magic behind this event did not occur in the minutes or hours leading up to this marquee game. Jordan had worked for years to develop the will to win and the skills to make the necessary shots even while ravaged by the flu. When he didn't make the varsity team his sophomore year of high school, he made a conscious decision to put forth colossal action to ensure that he would never experience the feeling of rejection or inadequacy again. He began putting in countless hours practicing to become the best of the best.

Colossal action is required to see remarkable transformations in your own life. Colossal action requires you to go beyond what's recommended by the status quo. Are you up to the challenge?

Change Your Environment

Have you ever been around someone who just seems to have that *spark*? They just have a passion and positivity that completely blinds them to the harsh realities of the world. When you're around this person, you feel like you have what it takes to take on the universe. These people emit an energy that is so powerful it elevates your own perspective. When these people become part of your environment, you start to see opportunities you were blind to previously.

Your environment—or the way you perceive it—is more important than you can even fathom. Real estate investor and author Robert Kiyosaki once stated, "One of the best ways to make changes in your life is to change your environment. This then changes you."

Your environment consists of the circumstances, objects, or conditions that surround you. To reference my previous story about my experiences in German class, my environment—in addition to my mindset—was highly instrumental to learning at a much faster pace. In this manufactured environment, we were told to only speak to the teacher in German. We would have German candy in class. Our tests would all be in German. For that brief time, our names would even be German names. The teacher did all of this to create an environment that was conducive to us learning this new language. For a little part of the day, we actually felt like German kids in class. All of this is to say your environment can be manufactured to create change in your life, no matter what change you want to bring about.

If your goal is to lose weight, you can alter your environment to make this goal more obtainable. Instead of a box of cookies on the counter, you can have a large bowl of fruit. The idea would be to create an environment that makes tough decisions easier, as well as helping you to form habits that are beneficial to your goals. Karl Benz created his own environment that allowed him to concentrate on his goals. He surrounded himself with others who had the same ambitions, along with the necessary tools to make his dreams a reality. His shop had all of the tools that he would need to work on the horseless carriage that would ultimately become the first practical automobile with an internal combustion engine.

> ***"One of the best ways to make changes in your life is to change your environment. This then changes you."***
>
> ***—ROBERT KIYOSAKI***

Do you have the tools necessary to implement the Business of Engineering *mindset*? Do you need to create an environment that is more conducive to creating the future you have envisioned for your personal and professional life? If not, now may be the time to shake things up. In the following sections, we will take a deep dive into the skills and attributes that make up the Business of Engineering. We will look at the skills that can be gleaned from a traditional four-year engineering degree and other skills that are most often learned outside the classroom.

CHAPTER 3 SUMMARY

MERCEDES-BENZ AND THE PATH TO SUCCESS

We've already discussed what the Business of Engineering means and why it will become so important to engineers and businesspeople. The next practical step was to understand how to implement these skills in your own personal and professional life. In order to learn these skills, you must take four steps. The four steps to jump-start your momentum are:

1. Change Your Mindset
2. Learn Voraciously
3. Take Colossal Action
4. Change Your Environment

Now, let's see how all of these engineering classes we took in college helped to prepare us for the business world.

"Engineering is the professional and systematic application of science to the efficient utilization of natural resources to produce wealth."

—THEODORE JESSE HOOVER

SECTION 2:

Engineering Curricula That Translate to Business

"The significant problems we face cannot be solved at the same level of thinking we were at when we created them."

—ALBERT EINSTEIN

CHAPTER 4:

IT'S JUST THE TIP OF THE ICEBERG— PROBLEM-SOLVING STRATEGIES THAT ACTUALLY WORK

THROUGHOUT THE ENGINEERING COLLEGE curriculum, there are many classes that are considered tough; then there are some classes that are downright grueling. For me, this class was thermodynamics. I went to school for civil engineering, so dynamic objects weren't my thing to begin with. As a civil engineer, if your objects were considerably dynamic (not stationary), you did something wrong, and you should probably exit the building or bridge you designed. For anyone who hasn't taken thermodynamics, it's the study of the relationship of heat with other forms of energy. This class was one of the most difficult classes I took in college.

Problem-solving was emphasized in all the classes we took in engineering school. Thermodynamics was no different. In this class, our problems consisted of enthalpy and internal energy, heat transfer through walls, and efficiencies of power systems. *Not* the most exciting stuff, in my opinion. I believe I scraped by with a passing grade in this class but not without a lot of work. With the projects I work on currently, I can safely say that none of those thermodynamic problems come up. Luckily, I don't regularly find the entropy change of an isobaric process. One thing this class did prepare me for was problem-solving in the real world.

The process of solving problems was so inherently ingrained in all of the classes in university that it would soon become second nature. Problem-solving wasn't just the process of finding the quickest answer to a quandary; it was finding the *best* possible solution. This meant you needed to factor all of the variables into the problem.

If you Google "problem-solving techniques," you will get millions of results. Many say there are four steps or even five steps to generate solutions. When solving a problem, I have found that six steps really need to take place in order to find the *most* viable solution:

1. Identify the problem.
2. Identify all of the variables.
3. Devise a plan of attack.
4. Execute the plan.
5. Evaluate the results.
6. Reevaluate with a contingency plan if the original plan generates inadequate results.

Next, I will describe each of the six steps. What's best about this method of problem-solving is that it can be utilized in engineering as well as in business!

Step 1. Identify the Problem

This may sound like a no-brainer, but first you need to know what the actual problem is. That is, you need to clearly articulate the actual problem your team is facing. For example, if your problem is that your consulting firm is not generating enough sales, is that the actual problem? Probably not. It would seem like a problem for your bottom line, but there's probably more to this issue than what you see on the surface. The actual problem could be that you haven't trained your employees to network properly, or you haven't satisfied your previous clients.

The problem is often hidden in the results that you see. To identify the problem, you must find the root of the problem. I know this may sound like something Yoda would say to a young Luke Skywalker, but stay with me. As the title describes, the tip of the iceberg is often all you can see of a certain problem. About 85 percent of the mass of an iceberg is below the water's surface. Your goal is to dive down to see the underlying problems below the surface. I often find this very difficult. One strategy I use to get to the root of a problem is to ask myself *why* at least five times. This technique has been attributed to Sakichi Toyoda, the famous founder of Toyota Industries and an engineer in his own right. He used this technique when he needed to decipher unique challenges he faced at his budding manufacturing plant. This technique is fairly simple and, at the same time, profound. This helps to dig deeper into a problem. Below is an example of what this internal discussion might sound like:

> We are having trouble obtaining city construction contracts.
>
> *Why?*
>
> We are located in a different city, and they seem to only work with firms within their city.
>
> *Why?*
>
> The firms within the city seem to have a closer relationship with the "movers and shakers" in the city.
>
> *Why?*
>
> They always attend city events and maintain contact with these officials.

By utilizing this "why" technique, it's easy to see that the firms located within the city receiving work are just more visible at municipal events. The real problem is that the competition is more noticeable than ours in city affairs.

Albert Einstein put it best: "If I had an hour to solve a problem, I'd spend fifty-five minutes thinking about the problem and five minutes thinking about solutions."

Here's another common anecdote on the subject: "A woodsman was once asked, 'What would you do if you had just five minutes to chop down a tree?' He answered, 'I would spend the first two and a half minutes sharpening my ax.' Let us take a few minutes to sharpen our perspective." Sometimes, the negative results seem like the problem, but there is a deeper problem. Your goal is to find it.

Step 2. Identify All of the Variables

This step might not be as apparent to most, but it was extremely important in engineering school. When working on a problem, I would write down all the relevant variables I could think of. Some were given in the narrative, some were not. You had to understand what variables were applicable to the problem, as well.

Consider the following problem: "If Jimmy and Tommy are six miles apart and Jimmy is walking at 3 mph and Tommy is walking at 2 mph, when will they meet?"

An additional variable that Tommy is eating a sandwich on his walk will not make any difference in solving the problem. I know this is a simplified and somewhat ridiculous example, but it gets the point across. The point is that all information you are given is not always relevant when solving certain problems.

The second step to solving problems is to locate all pertinent variables. That is, you need to search for all the clues you can find to help solve the problem. Let's go back to our example about the consulting firm not generating enough sales. After digging deeper, we find that the actual problem seems to be that the employees were never trained to network properly. If you're a business guru, you might think this is comical. Believe it or not, most engineers are not trained in social interaction for networking purposes. You identify that there are only one or two engineers who obtain contracts for the company. You also notice that 80 percent of the sales are coming from this 20 percent of your engineers (the old 80/20 rule, or Pareto Principle, for all of you purists). You've started to gather the facts on this problem you're having. But what do you do next?

Step 3. Come Up with a Plan of Attack

In engineering school, most of your memorization came from equations. This is very different from most other majors, who would memorize laws, theorems, and business models. These equations were tools for solving problems. We would load up our mental toolboxes with these tools to help us solve problems.

In a way, as business managers and leaders, you need to develop tools to solve problems as well. These tools or equations most likely won't be used to solve the entropy change of an isobaric process, but they will be extremely important when determining the correct courses of action in different situations.

Throughout your career, you will pick up tools as you go. You will learn most of these from your managers, and some you will just learn as situations arise. Another resource would be to read business books on management. You can literally pick the minds of some of the greatest business leaders throughout history by just picking up their books. Some of these authors include Jim Collins, Peter Drucker, and Simon Sinek. According to a Pew Research study, "Among all American adults, the average (mean) number of books read or listened to in the past year is 12, and the median (midpoint) number is five; in other words, half of adults read more than five books and half read fewer." This is a surprisingly small number. This is total books, which means it's not broken down into fiction or nonfiction. A majority of these books are most likely fiction books with little substance in regard to your professional development. I challenge you right now to pick up other books to learn as much as you can from the business titans of the past and present. The knowledge they present will give you many of the tools you will need to solve the problems you run into along your career path. Why stumble through problems when someone has already had those experiences and is willing to share the best way to solve them?

Step 4. Execute the Plan

This step is a given but is still worth mentioning. You would be surprised how many people don't take action on a problem and allow it to spiral out of control. You have to commit to the plan outlined in your brainstorming session, no matter how difficult it may be.

Some problems you will face in the business world won't be easy to address. They can often be problems you created. Although it can be tough to admit that you've created the problem, it's necessary if this is the case. The problem won't go away until it's confronted. Sometimes, this step is the toughest to actually follow through with. People often freeze up when it comes to making unpopular decisions, but it's imperative that you follow through.

Step 5. Evaluate the Results

Once you've taken action to solve the problem, the next step is to evaluate the results. Results can be measured against all sorts of metrics. Some of these metrics could be sales, profits, or production. You will need to define the metrics of your solution prior to the outcome. For example, let's say your company is producing some sort of widget and you believe production is lagging behind demand. You could quantify a success in solving this problem by increased widget production numbers. By the way, if this is your first exposure to the term widget, don't fret. The widget is often used as a stand-in for any product that is produced or manufactured. This term is often used in business management classes to define a product that you can quantify. A widget could be anything from a back scratcher to a toaster oven. The actual identity of the widget is not important but is used to help quantify outputs.

If you're an engineer reading this, you're probably thinking the same thing I thought when I first heard this term in my business classes. This widget is similar to the value x, or some other variable with a definite value, but is undefined in the problem. The widget, or x, can be anything you need it to be to measure the success of the solution you have provided for a certain problem. To validate your plan, a quantifiable outcome must be produced. Otherwise, you will have no means of determining the success of your solution. This is the key to defining true success in any endeavor you encounter. If your problem involves losing weight, you will not know if you are truly successful unless you measure your weight at intervals.

Step 6. Re-evaluate with a Contingency Plan If the Original Plan Generates Inadequate Results

Always have a plan B. Often, we have to resort to a plan B. When re-evaluating, you will start over at step 1. You will need to ensure that you've identified the problem correctly and that you've identified all of the variables. It's possible that you didn't identify these adequately in the beginning and thus did not produce meaningful results. See the graphic on the next page on the life cycle of problem-solving resolution for clarification.

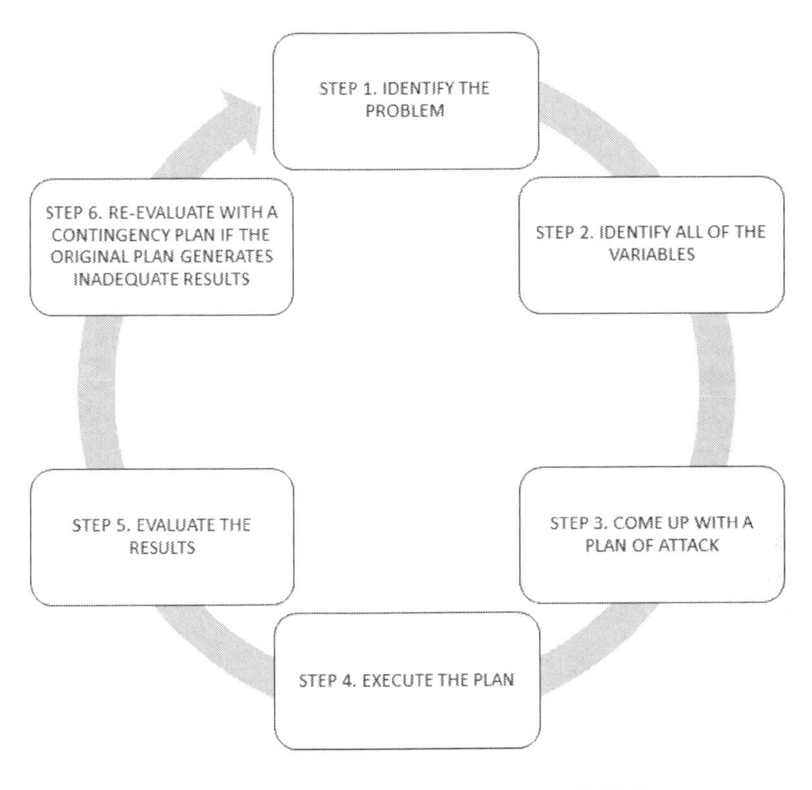

FIGURE 1. PROBLEM-SOLVING LIFE CYCLE

To give another example of using these steps in my professional career, I've included the following instance of problem-solving in a leadership role that I previously held. I recently finished a term as president of our company's young professionals group. Throughout my student and professional career, I've created and led many nonprofit groups. This leadership experience in particular was different and especially trying. The company I was working at had five different offices in two different states. These offices were geographically separate, with each one being about a two- to three-hour drive apart. This geographical separation made organization difficult for our relatively small group. We had a board of directors like any nonprofit, with representatives in each office. When writing the bylaws for the group, we assumed that this would be a sufficient representation of all the members.

For a while, it seemed like things were going great! Then we noticed member participation was starting to drop off. We thought maybe this was due to lack of interest or busy work schedules. After digging deeper, we found that the real problem was members didn't understand the full mission of the group (Step 1). Thus, they became disinterested and apathetic. As a board, we tried to understand how they might not have understood this mission. We talked to several of the members to get input (Step 2). After we gathered the facts, we had a long meeting as a board to discuss how we might reform the group to make it more meaningful for the members. The board decided we should start a newsletter to bring about a sense of community. We also made sure everyone had a copy of the bylaws and made every effort to communicate the mission of the group (Step 3). After brainstorming, the next step was to carry out our plan. We unveiled a newsletter outlining our purpose and added articles that highlighted events being organized by our young professionals group (Step 4). We found out many people enjoyed the newly minted newsletter and our other outreach efforts, but were they really helping? We approached several members and nonmembers to get their opinions. We wanted to find out if this outreach had helped our efforts to bring this group together (Step 5). After engaging the audience, feedback was mixed. We had to come up with a contingency plan and re-evaluate our plan to increase participation (Step 6). Examples like this happen every day in larger organizations. As a leader, you must dig deeper to find the most appropriate solutions to these problems.

Problem-solving is not just limited to engineers; it's a skill anyone can learn. In fact, many recruiters believe this skill is mandatory when looking for potential candidates to fill certain positions. According to a study of recruiters led by Wakefield Research, problem-solving was ranked as the most desirable soft skill needed when finding a candidate to fill management and leadership positions. Business endeavors require leaders to have the capacity to solve problems that others can't. This is how good businesses flourish. I challenge you to be inquisitive and dig just a little deeper into the exciting world of problem-solving. I guarantee you will be rewarded for solving problems that others simply cannot.

CHAPTER 4 SUMMARY

IT'S JUST THE TIP OF THE ICEBERG—PROBLEM-SOLVING STRATEGIES THAT ACTUALLY WORK

Don't worry, there isn't a thermodynamics quiz at the end of this chapter. We will just recap the steps necessary to really understand and truly solve the tough problems that plague businesses and engineers alike. Below are the steps for problem-solving:

1. Identify the problem.
2. Identify all of the variables.
3. Come up with a plan of attack.
4. Execute the plan.
5. Evaluate the results.
6. Re-evaluate with a contingency plan if original plan generates inadequate results.

Now that we understand the problem-solving process, let's delve into the impact of teamwork on solving these tough problems.

"A team isn't a bunch of kids out to win. A team is something you belong to, something you feel, something you have to earn."

—GORDON BOMBAY

CHAPTER 5:

PULLING TOGETHER ALL OF THE RIGHT PIECES FOR A HIGH-RISE PUZZLE

COACH BOMBAY STARES BLANKLY at the new hockey team he is preparing to coach. This team has never won a game. It's filled with an assortment of members. Some are athletic but most are not. Coach is obsessed with winning. This is going to be one of the toughest jobs he has ever had.

He quickly realizes this team cannot perform by forcing each individual member to become hockey juggernauts. He needs to leverage each member's unique talents for the advantage of the team. Coach has to create a strategy for the group to succeed as a team.

Over the course of the season, Coach Bombay starts to identify and utilize his players' individual skills. They start to win games, and Bombay becomes deeply entrenched in the culture of the team. They begin to believe in each other—so much so that they start to play like a team. They advance to the state championship. Locked in a penalty shoot-out for the ages, the team finds itself down to their last shot. The coach sends out the final player, who scores the goal! The team wins the state championship and the crowd goes wild!

If this doesn't sound familiar, you may not have been a child in the early 1990s. This is the storyline of the 1992 movie *The Mighty Ducks*. This movie may be a little silly, but it actually provides a lot of good insight into teamwork. This movie teaches about collaboration, communication, accountability, and individual strengths. We will delve into each one of these aspects in further detail. They are all extremely vital facets of teamwork.

As you can see, like most people, I started learning about the advantages of teamwork from a young age. This was reinforced even further in college. In college, I began to rely on teamwork even more with challenging assignments. Good teamwork hinges on the ability to collaborate.

Collaboration

Essentially, collaboration is considered teamwork but in more of an academic capacity. This skill is extremely valuable in engineering, as well as in business ventures.

Collaboration requires people to express ideas in a way that is easy for others to understand. This is often tough for engineers to do. Our vocabulary is generally technical in nature and includes jargon that others may find difficult to understand.

When you're collaborating with others, you want to make sure you're speaking their language—you don't want to use verbiage that is meant to confuse others. The goal of collaboration is to work together to find solutions to problems. You will need to leave your pride at the door for these sessions. We all like to think our idea is the best thing since sliced bread.

In order for collaboration to truly work, arrogance needs to be left out of the conversation. This doesn't mean you can't interject ideas or display your expertise; however, the goal of collaboration is to share ideas and hear others' points of view. It's hard to learn and understand someone else's point of view when you are always talking. As civil engineers, we are often the worst communicators. In a typical private development project, we work with architects, mechanical engineers, electrical engineers, plumbing engineers, owners, developers, and city officials. Everyone brings their own experience and expertise to the table. Clear communication is essential to make this design charrette work.

Communication

George Bernard Shaw once said, "The single biggest problem in communication is the illusion that it has taken place." This quote could not be more relevant to this discussion. Communication mediums have evolved over centuries, from smoke signals to instant messaging. Yet problems with communication still arise frequently.

For teamwork to truly work in a business context, communication needs to be clear and concise. Your team can consist of members of your organization or even stakeholders outside of your company. The problem that often arises is that everyone speaks their own language and communication seems to break down. It's your individual task to make sure people understand you. You can do this by asking if they understand your points. Or you can ask them to restate your request to confirm their understanding.

Undesirable outcomes can occur if communication breaks down among team members. Deadlines will be missed, and time and money will be lost. Never assume that your team members or subordinates understand your requests. Be sure to speak clearly and concisely. If e-mailing, keep it short and to the point. Sometimes, it might be better just to call, since text cannot convey inflection or other cues that may be important for understanding. People can often be offended by a text message or e-mail if it *sounds* arrogant or demanding, when in reality it was just a request.

Accountability

We've all had that one team member who never pulled their weight. They were always showing up late to meetings, forgetting portions of the required work, and making excuses for doing so. Take a minute to let your blood pressure lower. It's frustrating, but the only person you can really change is yourself. Accountability is essential for teamwork to be successful.

In a team, you are expected to use your personal strengths to better the team. Each member brings a different skill set and perspective. If even one of the members fails to pull their weight, the whole team crumbles. This is what I mean by saying accountability is essential for teamwork. Each member needs to be responsible for the success of the team.

How do you enforce accountability in your team projects? Accountability is about taking ownership for your work. You can't force your team members to be accountable, but as a leader you can reinforce the habit by becoming more accountable yourself. Provide a good example of this by taking ownership of outcomes that occurred based on your decisions. Never try to blame others for results that were undesirable. Once you have created a culture of accountability, it will become contagious.

Creating this culture of accountability will not be easy, but you can do a couple of things to foster it. Provide team members with individual goals and metrics. This will give each member incentive to pull their own weight. You can also give teammates greater responsibilities. This sounds counterintuitive because sometimes they can't handle the responsibilities given to them already, but when given greater responsibility, most people will gladly accept the challenge. This helps to promote ownership. Ownership and accountability go hand in hand. If members of the team feel like they have contributed significantly to a project, they will do their best to make it a success.

Individual Strengths

Individual strengths are what make teams unstoppable. When team members can fill one another's experience gaps, the team thrives. Each member should have their own distinct set of skills that brings value to the group. Experience brings ownership of the team goals and objectives. When a team member believes that their skill set is being utilized in the pursuit of the goals of the team, they become more engaged. We touched on this in the previous section. In the design charrettes I've been a part of, each member of the team has a unique skill to add to the equation.

A couple of years ago, the finishing touches of a ten-story office building were being made. This tower stood as homage to the late J. B. Hunt, founder of J. B. Hunt Transport Services. This structure was the tallest of its kind in northwest Arkansas. It still stands as a beacon of development and prosperity to the people there. However, most of the people in the area don't realize the teamwork that made this building a reality.

A little over three years ago, my previous firm got the civil engineering contract to handle the site work on this nine-acre campus. A local architecture firm handled the building's interior and exterior. A local structural engineering firm was responsible for the building interior support and foundation. A local mechanical engineering firm was brought on to design the utilities within this behemoth. After the team was assembled, they began collaborating. Each participant used their unique skill set to bring this structure to life. Each member had a stake in something that was much bigger than themselves.

After the design was completed and the construction company was selected, the discussions continued. The meetings were held at regular intervals. Each member of the team was called upon for their expertise. Sometimes, there were problems that came out of construction that affected multiple disciplines. In this case, the members would utilize their individual strengths to discuss the issues and how they would affect their designs. Then the team would discuss ways to remedy the problem without affecting other elements of the design. Each team member respected the experience brought by each of the other members. Although each member brought a unique skill set, the common goal of building the structure brought us together.

Sometimes, there were disagreements between disciplines. This regularly happens when you have several people with different values and backgrounds weighing in on a certain process or problem. Our group was very diverse, so disagreements arose often. It's what we did with these disagreements that made our group—and thus our project—successful. Collectively, this was the first high-rise structure this group had designed. With this designation came many unique requirements not normally seen with smaller projects. Most of the disagreements came from conflicts associated with different aspects of the mechanical systems with certain architectural and civil features. Each party involved in the conflict would discuss their need for the space or location, and then all parties in the construction meeting would give input on alternatives. This gave each designer a different viewpoint with which to see the conflict through.

After about eighteen months, the tower was completed. The structure was constructed on time, with few complications. The design team exhibited collaboration, communication, and accountability, as well as the team members' individual strengths. Collaboration was necessary to ensure that each piece of the puzzle would fit into place. Communication was utilized every step of the way. If effective communication was not in place, this story could have ended badly. Communication was necessary to transmit the ideas of each discipline in a language everyone could understand.

Accountability was paramount in this project. If a certain member or firm did not take accountability, the results could have been disastrous, or even deadly in the case of a ten-story office building. Each member maintained a sense of ownership of the project. The construction company even created apparel with the rendering of the tower for their tradespeople to wear. It gave them a sense of accomplishment. They became proud of the work they were doing and understood that it was much bigger than their individual tasks. Individual strengths were exploited often among disciplines to make this project a success. Each team member brought their own individual knowledge to the table. This allowed the team to identify conflict that might have otherwise been overlooked.

That office building still stands as a testament to economic success for the individuals in that region, but for me, it stands for much more. It stands as evidence of the effectiveness of interdisciplinary teamwork.

CHAPTER 5 SUMMARY

PULLING TOGETHER ALL OF THE RIGHT PIECES FOR A HIGH-RISE PUZZLE

Teamwork has always been a skill taught in Little League Baseball and Pee Wee football, but it seems to seldom get mentioned after that. Teamwork is an invaluable skill to learn and utilize throughout your career. These attributes of a successful team are visible in all high-performing groups:

1. Collaboration—thrives when personal agendas are set aside.
2. Communication—takes work to maintain but is truly essential for successful teamwork.
3. Accountability—creates a culture of accountability that starts with the team leaders.
4. Individual strengths—the true benefit for teams comes from the sum of all parts or the individual strengths of each of the members working together.

Next, let's delve into the details that make professionals wildly successful.

"It's the little details that are vital. Little things make big things happen."

—JOHN WOODEN

CHAPTER 6:

DOPAMINE AND THE STRUCTURAL BEAM PROBLEM

"THE DEVIL IS IN the details." How often have you heard someone at work or school say that? Throughout my engineering career, I have noticed that the details are usually what give people the most trouble. Throughout my time at the University of Arkansas, details often meant the difference between passing and failing. If you miss a certain detail in your construction plans, it could be very disastrous. The minutiae are extremely important, even in business matters. They show up in contracts, balance sheets, and expense reports.

When we were given problems in school, they were usually in the form of a word problem; we would sketch out all of the details on our papers. If it were a structural analysis problem, we would sketch out the beam or stationary object and then all the relevant information we could find hidden in the word problem. You would make note of certain things, like the length of the member, forces acting on the member, and certain structural properties necessary to answer the question or solve the problem. Many times, there were clues or details hidden deep in the problem that would not be entirely evident when initially reading the problem.

After solving many problems similar to this, I came to enjoy gathering the details in these problems. It was like solving a puzzle and having to find all the pieces. This was borderline exhilarating for most engineering students! If you're thinking, "Wow, glad I wasn't an engineering student," I wouldn't blame you at all. There might not have been quite as many parties attended by engineering students, but we still had a good time.

After working through many problems in various classes, the process of finding details and understanding which details were relevant became instinctive. Some details were given in the problem, and some were also implied. Implied details came through your knowledge of the type of problems being set forth along with your past experience of solving such problems. This skill would carry on into my career as an engineer.

In my experience as a civil engineer, I've learned that details can mean the difference between life and death for the citizens who utilize the designs that were constructed. I briefly described civil engineering earlier, but it's the engineering of structures that facilitate public works. We take our job very seriously because our end user is the public. One of the professional societies for civil engineering, ASCE (American Society of Civil Engineers), states that an engineer's fundamental responsibility is to "hold paramount the safety, health, and welfare of the public and shall strive to comply with the principles of sustainable development in the performance of their professional duties." In order to truly uphold this canon, it's necessary to pay attention to the details.

Civil engineers produce drawings or blueprints of structures they have designed. These blueprints must lay out all of the relevant and important details needed for the contractor to construct the structure or development. The plans usually depict materials for construction, dimensions, and other relevant information. An engineer's job is to successfully communicate to the contractor the designs we have calculated. If the details of the construction are effectively relayed to the contractor, they should be able to successfully build the desired structure.

Unfortunately, plans are rarely, if ever, perfect. Human error is just a part of life. We can do our best to try to prevent mistakes, but we can only stare at the details for so long before our production slows or grinds to a halt. It's just not feasible to review plans for longer than it takes to produce them. The best we can do is to reduce mistakes when producing the plans. There are errors in everything humans produce. Some people argue that most of the work is done by or with computers, but computers are only as accurate as the humans who program or utilize them. "Garbage in, garbage out," as the old saying goes. So how do we prevent human errors from being introduced into our plans, spreadsheets, and reports?

After trying several techniques in my own day-to-day activities, I found five techniques that have helped me become more aware of the details in my work. The majority of these techniques boil down to being disciplined. Having discipline in your work and personal life allows you to become more productive. The five habits that I have found to be most helpful when trying to limit miscalculations are the following:

1. Planning your projects in advance
2. Creating a project schedule
3. Eliminating or reducing distractions
4. Having a trusted colleague back-check your work
5. Maintaining physical and mental health

Seems pretty simple, right? It takes a lot of discipline to make these techniques become habit, but once you do, I promise your productivity and accuracy will increase. In the next section, we will discuss these habits and how they might be introduced into your everyday work life.

Planning Your Projects in Advance

Have you ever had a day when things just didn't go right? You wake up late, throw your alarm on the floor in haste, and jump in the shower. You run out the door with a splash of coffee and piece of toast, and you're on your way to the office. Next thing you know, you're stuck answering e-mails for the first hour of work, and don't forget to adjust your fantasy football lineup, either.

Your day was shot before it even started. There was no schedule or plan to your day. Many people get stressed doing urgent, nonessential tasks because they haven't produced a schedule for their projects.

There's nothing magical about planning your day. Here's a very generic timeline of how you might schedule your time. Maybe you schedule the first two hours of work to tackle your most daunting project, then you move on to your next big hurdle. Eventually, you will work your way down to the tasks that don't seem as overwhelming. We can argue that our brains function more clearly in the morning, but the main reason for getting it out of the way early is to reduce the distraction of this problem or project. If it's completed, you don't have to think about it anymore. This equates to less distraction on your next item.

To utilize this technique fully, you must be able to say "no" to time wasters such as Facebook, Twitter, and surfing the Internet. This habit can take a lot of discipline, but I guarantee that once implemented, your life will be a lot less stressful.

Creating a Project Schedule

This is a skill I picked up after graduating from college. The projects I currently work on can take two to three months to be fully permitted. In school, our projects lasted about two to three weeks. This wasn't really a long enough time frame to lose track of the project. At any given point, I probably juggle three to five different projects at a time. This wasn't the case in college, either. With these longer timelines and diverse projects, scheduling became imperative.

This scheduling routine requires a little up-front use of your time, but it's well worth it. Once you've created a template for your projects, it becomes much easier. Our consulting firms' projects take place in various cities and encompass different types of developments. It's imperative to create schedules for these projects to stay on the proper timetable. When the timeline is in place, the project becomes much smoother. Everyone on your team knows when they're expected to have their pieces finished. This tactic minimizes errors because no one is rushing around at the last minute to complete a project. We also make sure we have scheduled in opportunities for peer review that might not have taken place otherwise. I will talk about the review process further, but this is an integral piece of the detail puzzle.

Eliminating or Reducing Distractions

Welcome to the twenty-first century. Distractions are literally everywhere you look. How can we avoid getting caught up in distractions when everything we own seems to buzz, beep, or flash at us? This habit may be the hardest, but I guarantee it will be the most life-changing.

Distractions come in the form of e-mails, texts, phone calls, coworkers, and other variables in your environment. In a Michigan State University study, three hundred participants were given a task to complete consisting of producing a certain sequence on a computer. This chore mimicked the usual work tasks of a majority of office workers. Then they produced distractions at various intervals. They found out that it only took a distraction of less than three seconds to result in twice the errors as their original effort. **Less than three seconds.** That's similar to just glancing at your phone to read a text message. So why would this very short distraction cause so many errors? According to the lead researcher, Erik Altmann, "The answer is that the participants had to shift their attention from one task to another. Even momentary interruptions can seem jarring when they occur during a process that takes considerable thought."

As I'm sitting here typing this book, my phone's notifications are constantly competing for my attention. So how do we take our attention back from all the technology that now surrounds us? One strategy is to simply turn off the notifications on our devices. This might sound too intuitive, but without that little pop-up to check an e-mail, or the ding of your phone notifying you that you have a message, I can guarantee that you would not check your messages nearly as regularly. That "little" notification is intoxicating. It sends a rush of dopamine to your brain, like that of drugs or sex. The unpredictability of the message triggers your brain to produce dopamine. You can actually become addicted to these alerts. Scary, right? I suggest picking two to three times a day to check e-mails and respond to calls or messages. You may miss a few "important" e-mails now and then, but your productivity will increase exponentially with just this small shift.

In college, there were various study groups I would meet with to trudge through thermodynamics homework or study for the next big test. These groups were great when you didn't understand a topic, because you could often get someone else's take on a subject. They were not so great when you had already grasped the concepts, because of the distraction caused by your peers.

This takes place in corporate offices throughout the business world. Many distractions are in the form of questions or requests from coworkers, managers, or clients. How do you eliminate these distractions without abolishing the company's cultural norms and still maintain relationships? The key is communicating your needs up front. Tell your roaming colleague—who's often looking for a distraction themselves—that the best way to ask you a question is to send an e-mail, and you will get back to them. You may lose a few friends at first, but you're becoming more productive and reclaiming your time. You shouldn't feel bad about this. They will eventually respect your requests and start to make it a habit of sending an e-mail first. You will then answer e-mails when you originally planned to. This will significantly cut back on distractions at work.

With all the distractions these days, it's a wonder anything gets done. You now have the advice and tools to reclaim your day from the power of disruptions. All of these distraction-elimination techniques require discipline. Discipline to turn off notifications, discipline to confront diverting coworkers, and discipline to block certain websites. The next technique that allows you to pay attention to all the details is to disrupt another coworker. Ironic, right?

> ***"An addiction to distraction is the end of your creative production."***
>
> ***—ROBIN SHARMA***

Having a Trusted Colleague Back-Check Your Work

What if I told you that you could have a stellar classmate check your test before turning it in to the teacher? In college, this would have been great (especially in thermodynamics). I can guarantee you that some of my errors would have been caught if I'd done this. I was never a bad student, but having another set of eyes on your work can catch errors that you might not have found otherwise.

This technique can be utilized regardless of your profession or degree. It never hurts to have a trusted colleague back-check your reports, designs, or calculations. Almost every report I've written in my career has been reviewed for quality control. Like I said in the beginning of this chapter, mistakes are part of being human. Your odds of making mistakes decrease whenever you have someone unfamiliar with the project review it. There are many times when I've looked at a set of plans so many times that my brain was completely numb. The idea is to give these plans to someone who is unacquainted with them. They have a greater possibility of finding errors because it's new to them. I would highly recommend that you have someone of equal or greater experience back-check your work.

Maintaining Mental and Physical Health

This next habit might not make much sense initially, but once we delve deeper, you will understand its importance. When you look good, you feel good. The same rings true for your performance at work. Studies show that a healthy worker is a productive worker. A study from the Health Enhancement Research Organization (HERO), the Healthways Center for Health Research, and Brigham Young University (BYU) reveals that employees who maintained healthy diets throughout their day were 25 percent more productive than their peers who didn't. The same is true for those who maintained thirty minutes of exercise at least three times a week. The participants who maintained an exercise regimen were shown to be fifteen percent more productive. I try to exercise at a nearby gym at least three times a week at lunch. If you can't be guaranteed a lunch break to work out, mornings are a great time to get it done. That's usually the only time I can guarantee time without interruptions. Although this is not a book on health and nutrition, there are many other great resources on the subject. A few of my favorite nutrition and exercise books are *Whole: Rethinking the Science of Nutrition* by T. Colin Campbell, *Strength Training Anatomy* by Frederic Delavier, and *Salt Sugar Fat: How the Food Giants Hooked Us* by Michael Moss. In addition to a healthy diet and exercise, sleep is paramount.

According to a 2008 National Sleep Foundation poll, one-third of all Americans say that daytime drowsiness affects their daily work activities at least a few days each week.

I admit that I probably don't get the recommended amount of sleep (seven to nine hours for adults). It's hard to find time in the day to reach all of your goals and sleep for nine hours. However, there are ways to get more sleep. I've listed a couple of habits on the next page that can help you regain time for sleep and be more productive and alert. Like every other habit listed previously, these habits require discipline, but like the rest, they can be very rewarding when followed:

❑ **Manage your time well.**
  - This goes without saying, but the reason you might not be getting enough sleep is that you aren't managing your tasks efficiently throughout the day.

❑ **Set hours that you will not work.**
  - This habit goes along with the previous item. Set hours—seven to nine p.m., for example—that you absolutely will not work.

❑ **Reduce stress.**
  - Stress at work can often cause you to stay up late. Try adding some stress-busting techniques like meditation or yoga to your schedule. I've worked yoga into my schedule on occasion, and it does wonders for my quality of sleep.

❑ **Establish a bedtime.**
  - I know you're not ten years old anymore but setting a bedtime (and sticking to it) can be liberating. Having a schedule for your bedtime allows you to schedule your evenings for maximum productivity, so you go to sleep on time.

These habits are just some examples of ways you can regain your precious sleep time. I challenge you to come up with some more ways to get your seven to nine hours of recommended sleep. I guarantee you will notice the benefits of this habit soon after you start.

As you can see, there are many ways to improve your accuracy and attention to detail in the workplace. Once you have successfully introduced these habits into your own career, I guarantee you will notice small improvements in your ability to catch details that might have otherwise escaped you, and your productivity and mindset will improve.

CHAPTER 6 SUMMARY

DOPAMINE AND THE STRUCTURAL BEAM PROBLEM

Details are everywhere. They can often mean the difference between a project or campaign that's successful and one that's not. Although we can't be perfect, we can get a little closer by following the habits below:

1. Plan your projects in advance.
2. Create a project schedule.
3. Eliminate or reduce distractions.
4. Have a trusted colleague back-check your work.
5. Maintain physical and mental health.

These habits demand discipline, but the rewards can be limitless. Next, we will delve into the wild world of creativity and how to regain this long-lost trait.

"You see things,
and you say, 'Why?'
But I dream things that
never were, and I say,
'Why not?'"

—GEORGE BERNARD SHAW

CHAPTER 7:

INGENUITY AND THE ENGINEER—HOW TO HARNESS CREATIVITY

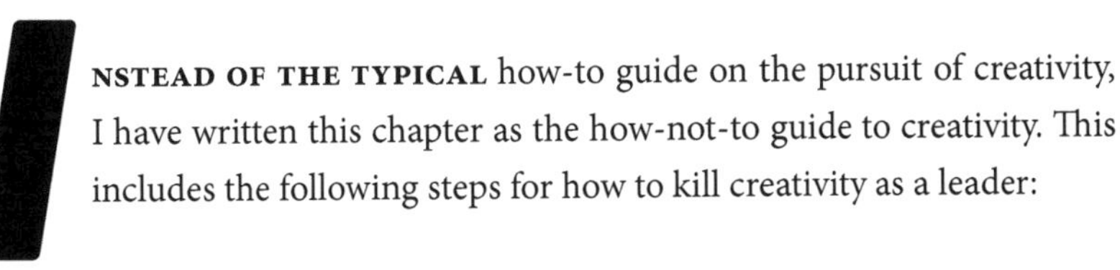

INSTEAD OF THE TYPICAL how-to guide on the pursuit of creativity, I have written this chapter as the how-not-to guide to creativity. This includes the following steps for how to kill creativity as a leader:

1. Always pretend to know more than everybody in the room.
2. Ensure that highly qualified people do extremely routine work for extended periods of time.
3. Create physical and metaphorical barriers between departments.
4. Don't speak empathetically to subordinates, except when announcing increased goals, reduced deadlines, and squeezed cost restraints.
5. Schedule multiple unnecessary meetings.

In the following pages, these creativity killers will be discussed in detail, with examples I've seen in my career and how they can be prevented. So let's stop wasting time and jump right into the list!

Always Pretend to Know More Than Everybody in the Room

This is number 1 on the list for good reason. Experience in a certain job or field can be a blessing or a curse. In the book *Rookie Smarts: Why Learning Beats Knowing in the New Game of Work*, Liz Wiseman states, "When the world is changing quickly, experience can become a curse, trapping us in old ways of doing and knowing, while inexperience can be a blessing, freeing us to improvise and adapt quickly to changing circumstances." I have found that over my career, my inexperienced peers seem to come up with more creative solutions to problems than my more experienced cohorts. Why is this?

Inexperience breeds curiosity. When you're inexperienced, you don't get stuck in the trap of doing something the way it's always been done. As a newcomer, you don't know everything, but you know those who do. You begin to seek out information from those who are more knowledgeable than yourself. The novice then begins to compile all of their resources before making a decision. They don't let ego get in the way of seeking out helpful information. The greenhorn often seeks out knowledge from anyone located along the supply chain, including shareholders, producers, suppliers, and even end users. This glut of data often gives the tenderfoot a unique perspective on a problem or process. They don't pretend to know everything, because they simply don't. This is often seen as a disadvantage, but when harnessed correctly, it can be a giant benefit.

An experienced cohort will often rely on the processes that have sustained satisfactory results in the past with little input from others. When you're experienced, you don't question the status quo because it becomes uncomfortable. You're satisfied with the old way of doing things and don't seek additional outside help because it's deemed unnecessary and, in many cases, uncomfortable. When you're a veteran, you often believe you know everything there is to know about your area of expertise. Unfortunately, this ego stymies creativity. It eliminates collaboration and deters inspiration. By challenging the status quo, peers can stimulate creativity.

This phenomenon of creativity in newcomers is everywhere you look. I see it often in my workplace. I've even been guilty of believing I know it all. I had become comfortable with my processes and skill sets when an even younger colleague shook my perception of certain solutions. I had become trapped in the veteran mindset. I wasn't thinking like a greenhorn anymore. There have been many times when I have gotten stuck in the process of doing certain design aspects of my job and wouldn't take the time to step back and look at my tasks from a thirty-thousand-foot view. Taking the time to really examine your tasks from a fresh point of view is exciting—and even a little liberating.

As a young engineer, I see this as a distinct advantage over more senior engineers. Although they may have the wisdom and experience of years of problem-solving, they begin to form patterns of solving these problems. If you're a new engineer, like I consider myself to be, use this to your advantage. Continue to think outside the proverbial box. Always seek multiple opinions when tackling a problem. If you're a more seasoned engineer or businessperson, play pretend! It may sound elementary, but when you set aside your old processes, you can open up a new environment of creativity. Go back to a time when you just started a job. How did this feel? Overwhelming? Exciting? You can spark your creativity by reenacting your experience. Before jumping right into a typical solution protocol, ask yourself, "Is this really the best way to do this? Is there another way I could do this?" Be sure to ask questions and enjoy the often-uncomfortable thrill of seeking creativity.

Ensure That Highly Qualified People Do Extremely Routine Work for Extended Periods of Time

This next rule for killing creativity is similar to the last. It's more relevant to someone in a leadership position, though. This "routine" creates a rut that can be hard to get out of. If you have driven a vehicle into a rut, you know that it can be difficult—almost impossible—to get out of. Picture yourself driving down an old dirt road. The recent spring rain has made the once-hardened thoroughfare into a spongy path. Your truck rocks as the tires yield to the perfectly formed tire ruts. Mud begins to fly everywhere, spattering your vehicle with sludge. You jerk the steering wheel from side to side, but there's no getting out of these perfectly shaped tire traps. The vehicles that have traveled this same trail over and over have formed a path in which your vehicle almost feels compelled to travel. This is a near-perfect metaphor for the highly qualified professional who becomes trapped by the dogma of "this is the way we always did it." When these dogmatic professionals get stuck in this mindset, it becomes routine. This predictable problem-solving route becomes a barrier to creative thinking. If you feel like you have gotten your wheels stuck in this rut, try to ask questions to your younger, less experienced cohorts. Become curious about the process. Try to look at your process from thirty thousand feet and see if it's still relevant. Are there ways to improve the process? Do you just need to shake up the routine, or does this work even interest you anymore? People who get stuck in routines at work are often unsatisfied with their jobs. If this sounds like you, you might look into other options that provide you with the ability to be more creative.

Create Physical and Metaphorical Barriers Between Departments

A lack of collaboration can kill any hope of creativity in an organization. Oliver Wendell Holmes said it best: "Many ideas grow better when transplanted into another mind than the one where they sprang up." We discussed collaboration briefly in Chapter 5, but it's also important when coming up with creative ideas. I often will find that when I discuss a problem with professionals from other disciplines, such as architects, contractors, and mechanical engineers, potential solutions may impact their designs. If we never discussed the potential problem and its solution, I may have designed a noisy transformer where they intended to have a peaceful park next to a building. Communication and collaboration are paramount to ensuring a project's success and can give birth to some creative solutions.

Don't Speak Empathetically to Subordinates. Only Remind Them about Unrealistic Goals, Reduced Deadlines, and Squeezed Cost Constraints.

As a leader, which I'm guessing you are if you're reading this book, this should hit home for you. If you're an engineer reading this book, the word empathetically may be scary. A few years ago, I would have agreed with you. We discuss empathy in the relationship-building chapter, but it is really important throughout the book. As my wife will certainly tell you, I am not the best at sharing my feelings. This notion goes much deeper than that, though.

In regard to creativity, empathy can be used to dig deeper into the methodologies used by a subordinate or peer to come up with a certain creative idea. The idea is to approach their concepts free of judgment. Once you make it a "judgment-free zone" (perfect saying taken from Planet Fitness), they will be more apt to share creative ideas. Fear of sounding uneducated, ignorant, or even oblivious keeps many people from sharing their ideas. Your job is to encourage such input regardless of how it will sound. Then you can really dig in and examine other thought processes. This ideology will help build rapport with your subordinates and colleagues. They will trust you not to ridicule their ideas and in turn will be freer to share them.

Schedule Multiple Unnecessary Meetings

Nothing kills productivity and creativity more than an absurd number of meetings without any agenda. These meetings are always at the "wrong" time for me. I don't know about you, but what kills my productivity and creativity the most is having to stop what I'm doing and pile into a room with several other people and discuss (insert arbitrary topic) for an hour without any clear outline. These types of meetings are usually a huge waste of time for the people attending them. This ritual or business sacrament often wanders off topic without clear direction. But meetings aren't always bad—when they're executed properly.

When you have to schedule these meetings yourself, remember to prepare an agenda and stick to it. It is the least you can do to assure attendees there's a reason and time limit for this meeting. This will help keep the attendees focused and interactive, knowing that there is only a certain amount of time allotted for certain discussions. Ensure that everyone included in the meeting really does need to be there. I've been stuck in meetings before when there was no real reason for me to be there. I'm sure you've had the same experience. I have been stuck in meetings with architects when, as an engineer, I don't normally have much to say on the differences between stucco and brick façades.

When you're actually the ringmaster for the meeting, be sure to start the meeting on time and not a minute late. Don't wait on the usual latecomers. This will show that you have an affinity for promptness. Nobody wants to show up late in the middle of a meeting. It's embarrassing. After a few meetings, peers will respect your requirement for punctuality. Before you schedule your next meeting, ask yourself if this meeting is truly necessary. The costs to productivity and creativity may far outweigh the need for such a procedural norm.

As consumption of resources continues to increase, creativity will be even more important to ensure longevity for future generations. The ability to generate and implement creative solutions can ultimately decide the future, especially when it comes to dealing with some of the world's biggest challenges like failing infrastructure and global warming. As cheesy as this may sound, we need to embrace creativity as a necessary skill set and not let it fall by the wayside with other skills that were nurtured in our youth, such as imagination and curiosity.

Imagine if we all included creativity in our day-to-day activities. Wouldn't your workday have an almost whimsical feel to it? What if we were free to let our minds explore various solutions to problems—without ridicule. This is all very possible. Just like any habit or skill in this book, it requires a shift away from the belief of "we always do things this way." This dogma traps organizations and individuals and shackles them from producing truly extraordinary results. I challenge you to take off the proverbial blinders and look at your projects with a new zest. Look at them with fresh eyes and see what you're missing "in plain sight."

CHAPTER 7 SUMMARY

INGENUITY AND THE ENGINEER—HOW TO HARNESS CREATIVITY

The word "creativity" has a whimsical stigma that some professionals consider frivolous. Let's challenge that notion and dare others to do the same. If you want to **kill** creativity among your team members, these habits will certainly do that:

1. Always pretend to know more than everybody in the room.
2. Ensure that highly qualified people do extremely routine work for extended periods of time.
3. Create physical and metaphorical barriers between departments.
4. Don't speak empathetically to subordinates, except when announcing unrealistic goals, reduced deadlines, and squeezed cost constraints.
5. Schedule multiple unnecessary meetings.

To cultivate creativity, do the absolute opposite! Next, we will delve into deadlines, which can support creativity if handled correctly.

"I've learned over decades of building that a deadline is a potent tool for problem-solving."

—ADAM SAVAGE

CHAPTER 8:

THE IMPACT OF THE "ZERO HOUR" AND UTILIZING CRUNCH TIME

THE FAMILIAR WORD "DEADLINE" had ominous beginnings when first coined back in the late 1800s. As with many business words we use today, this word stemmed from war. In Andersonville, Georgia, a prison camp for Union prisoners of war was constructed. This camp came to be known as Camp Sumter, and it was the largest Confederate prison camp constructed during the Civil War.

The walls were constructed of pine logs of staggering height. Along each wall stood lookout towers for the Confederate soldiers to keep an eye on the Union prisoners. Located approximately nineteen feet from the wall was a small fencerow. This fence was far enough from the wall to prevent someone from scaling over or tunneling under it. This smaller fence was considered the "deadline." If a Union prisoner crossed this literal deadline, they were shot by the Confederate soldiers keeping watch in the towers.

Your project deadline doesn't sound so bad now, does it? This may be why deadlines are always imposed with a negative connotation. Deadlines are a necessary evil, whether self-imposed or not. In the following pages, we will discuss the benefits of deadlines, the consequences of deadlines, and how to make deadlines work for us.

Deadlines come in all shapes and sizes. They can be imposed by yourself, a client, or even a regulatory department. But few people truly realize the power of deadlines. Deadlines, although stressful, can create powerful goals that help streamline the problem-solving process. Problem-solving, as outlined in Chapter 3, can be a long and arduous process. We discussed previously that you want to follow a process, but what if your timeline is cut in half? Do you think you could still produce meaningful results?

Surprisingly, deadlines can be a powerful tool in productivity. Deadlines are simply a timeline for finishing a project or task. When you assign a deadline to a task, you understand when it needs to be finished. This allows you to prioritize certain projects. If you don't have a deadline on a certain task, it tends to slip further back into the proverbial pile. This is where the power of deadlines comes in! When you create deadlines for certain tasks, you are prioritizing them. This allows you to stay focused on a certain project because you know that time is of the essence.

We have all done it before. We have put off projects or research papers until a day or two before they were due. We were given deadlines a month or two in advance and yet we still struggled to complete the task. Why does this happen? This behavior can be attributed to procrastination, and we're all guilty of it. Procrastination seems to creep in whenever you believe you have ample time to complete something. In a study done at the University of Chicago, a group of researchers studied the impact of deadlines on consumer action. According to the authors of the study, "the way consumers think about the future influences whether they get started on tasks. In particular, if the deadline for a task is categorized as being similar to the present, they are more likely to initiate the task." In other words, when a deadline seems more immediate, the consumer will act on it. To encourage this type of action on your own, short, self-imposed deadlines are ideal.

You'll find that you really don't need as much time to do certain tasks as you previously thought. Shorter deadlines will help you to cut out the typical distractions. I have found that giving myself shorter deadlines is more productive than allowing myself longer ones and has helped me to increase my production. According to Parkinson's Law, "work expands to fill the time available for its completion." That is, people will find ways to fill in the time to complete a task, but it might not be related to that specific task. This law can be seen in many corporations around the country. Without clear focus on the tasks at hand, many employees can get distracted by the wonders of Facebook, Instagram, and other Internet disruptions. Focus is created by strict deadlines. If the deadlines are easily within reach, like the research papers that were due months in advance, people will find ways to fill the time—and usually not in ways that will be conducive to finishing the paper.

How many times have you been given an "outrageous" deadline on a project that didn't seem possible? Maybe it was a couple days to turn around a project that would normally take a week or two to complete. Or it was a homework assignment that seemed impossible to finish before the next class. We've all experienced these so-called miracle assignment completions. How did you feel after these assignments were turned in? Exhausted, relieved, or both? How about self-confident? When I finish these "impossible" tasks, I often feel like I'm on top of the world, like the scene in *Rocky II* where Rocky runs to the top of the steps of the Philadelphia Museum of Art. The iconic scene shows Rocky running through the streets of Philadelphia with throngs of cheering kids and adults following him, shouting his name. He then makes his way up the steps of the art museum while the song "Gonna Fly Now" plays in the background. At the top of the steps, he celebrates with the many followers who made the "impossible" journey with him. This is a dramatization of a seemingly dissimilar event, but when I finish these tasks, I become confident that I can do anything. Beating an impossible deadline gives me a confidence and enthusiasm that I can transfer to the next project. I believe you will find the same thing happens when you complete your next challenging assignment. The build-up of adrenaline created by the prospect of an "impossible" deadline fosters a sensation of invincibility once it's completed.

The clock is ticking—10, 9, 8. The secret agent is frantically searching for a solution to the very tight deadline in front of him. The ticking bomb will detonate and kill many of the unsuspecting civilians in the building if he does not act quickly. The tight space the bomb is in will create a vacuum that will cause massive damage to the building and its occupants. 7, 6, 5. The agent's body is flooded with adrenaline. He is faced with the typical "fight or flight" reactions. He can either jump out of the building and save himself, or he can solve this problem with the possibility of sacrifice. 4, 3, 2. At the last second, the hero grabs the bomb and hurls it out the window toward the open sky. The energy of the bomb dissipates in the night sky, and everyone in the building is saved.

Tight deadlines also increase creativity in a situation. In the example above, the hero has a very tight deadline to process all the outcomes of the situation. He can either run away at the expense of the unsuspecting victims, or he can come up with a solution to solve the problem in front of him. These heroes will often come up with solutions that aren't inherently apparent. These highly dramatized situations are not self-imposed, but they share many of the same characteristics. When faced with a seemingly impossible deadline, tap into your creativity to come up with overlooked or unexamined solutions. These expedited deadlines can force you to think outside the box. This often-overused cliché is underutilized in corporate America. It's easy to become complacent with our typical tasks and deadlines. We create processes to complete certain tasks in certain amounts of time with certain results. Tight deadlines help to shake things up.

So, how do we implement deadlines to help us achieve our full potential? There are a few key steps in making successful self-imposed deadlines. These guidelines are not revolutionary or groundbreaking but when used properly will enable you to quickly tap into your creative side:

1. Set a clear and defined deadline.
2. Instill a sense of urgency.
3. Create a plan of action.
4. Utilize a reward system.

Let's get right into it.

Set a Clear and Defined Deadline

How often have you started a project only to sock it away for a while until it becomes really important? We often see the deadline as a distant and intangible mark on our calendar. When you start your self-imposed deadline journey, you will need to quantify your target. Let's say you want to lose weight by your ten-year high school reunion. That sounds great, right? We all have a little bit of extra weight we'd like to lose, but this is a hollow deadline. Sure, you have a "due date" by mentioning that it must be completed by your high school reunion. What constitutes a successful outcome for this task? Is it losing ten pounds, twenty pounds? I think you get where I'm going with this. Without a clearly defined deadline, your mind will create its own definition of success. Unfortunately, it's often not truly aligned with your goal.

When setting a deadline, be sure to enumerate the measures of success, as well as define your duration. These are the keys to a fruitful self-imposed deadline. When writing this book, I gave myself a clearly defined deadline. I would then produce mini-deadlines each day. I would say that each day I wanted to write for at least one hour and gave myself a specific goal of two pages a day. Now, these pages weren't always finished masterpieces, but they helped me to reach my ultimate goal.

Instill a Sense of Urgency

We don't always take ourselves seriously, unfortunately. We believe our threats are often hollow, unlike those of our bosses and supervisors. The way to combat this feeling is to give yourself far less time to complete a task than you previously thought possible. Give yourself days instead of weeks to finish a task. You'll be surprised to find out that you usually have way more time than you need to complete certain tasks.

In 2005, Apple was in a bind to find the next big thing that would solidify the company at the top of the personal technology realm. Steve Jobs gave the engineers of the original iPhone a couple of weeks to decide what the future of software was for the company. He made threats to fire them if they could not complete this task. Needless to say, this strategy produced a high sense of urgency for these talented engineers. Out of this great sense of urgency emerged the iPhone. Today, market share of the iPhone hovers around 20 percent of all smartphones. This Jobs-imposed deadline forced the engineers to think outside of the box and gave them a sense of urgency to complete the task.

When we give ourselves more than enough time to complete a task, we begin to stress about it as other commitments pile up. What if we decided to shorten our deadlines and actually focus on a project or task? Instead of letting the longer timelines hang over our heads, we tackle them head-on and focus solely on that one task. I believe you would encounter far fewer mistakes in the process, and your stress level would plummet. You would not be jumping around from project to project and creating a sense of "busyness." So, next time your boss says you have three weeks to complete a project, tell them you will have it to them in one. Like the actor disarming the bomb with the countdown timer ticking to zero, I believe you can find the right solution in less time than you thought possible.

Create a Plan of Action

You can't meet these expedited deadlines without a clear plan of action. So, what is a plan exactly? My favorite definition is from Merriam-Webster's dictionary. A plan is defined as "a method for achieving an end." This simple definition is the key to understanding the need for such a plan of action. You will create a method for achieving some sort of mission.

A plan of action doesn't need to be highly detailed nor does it have to be extremely complicated. It just needs to include small, measurable milestones that will signal that you are on the right path to reaching your goals. The key to these milestones along your plan of action is that they need to be measurable. People can become addicted to progress. I admit that when I have reached several goals on the way to my larger, more universal goal, I get pretty pumped. The idea that you are moving toward your goal gives you a sense of achievement. This triggers a bit of dopamine in your brain and gets you ready to tackle the next hurdle. When you have a plan of action, you understand what the next obstacle will be, so there's no need to slow down and look around for the next target.

Utilize a Reward System

People love rewards, plain and simple. Whether it's a child who gets an allowance for taking out the trash or an adult who gets a raise at work, we all love receiving rewards. What better way to light a fire under yourself when working on tasks than a tangible reward? Obviously, achieving your goals will itself be a reward, but why not get something that you really want? On the hilarious *Parks and Recreation* sitcom, some of the main characters have a "treat yo'self" day devoted to rewarding themselves for no real reason. They buy outrageous clothes, get pampered, and eat whatever they want. Now, I'm not saying this is the best idea, but you deserve to "treat yo'self" if you tackle an arduous deadline! This could be something as small as getting a massage for finishing a report or going on a luxurious mini-vacation with your spouse for finishing your master's program. By setting up these rewards, you'll be more enticed to keep focus and momentum when the going gets tough.

Deadlines can either be your best friend or your worst enemy. Like most things in life, all it takes is a shift in mindset to see things from a different point of view. What if you made this all-too-important mindset shift from disgust to the utilization of this powerful productivity tool? What would happen if you decided to exploit the often-misunderstood deadline to increase your productivity and creativity? I think you would understand the full potential that comes from shorter targets and constraining time limits. Give it a shot, and get ready to reap the rewards of ever-increasing efficiency.

CHAPTER 8 SUMMARY

THE IMPACT OF THE "ZERO HOUR" AND UTILIZING CRUNCH TIME

The word *deadline* sounds ominous and inauspicious. However, there is a lighter side to deadlines, as you have discovered in this chapter. Below are a few of the points of interest from this chapter:

1. Deadlines, although sometimes stressful, can become great self-imposed goals that increase productivity.

2. Deadlines can actually streamline the problem-solving process.

3. Non-self-imposed deadlines can have expensive consequences.

4. There are many ways to handle deadlines in order to make them work to our advantage.

See, deadlines aren't all bad. In fact, they can be utilized to work to our advantage. Next time you sit down to start a task, try giving yourself deadlines to hit certain milestones. I guarantee you will be satisfied with the results.

"Nothing in the world can take the place of persistence. Talent will not; nothing is more common than unsuccessful men with talent. Genius will not; unrewarded genius is almost a proverb. Education will not; the world is full of educated derelicts. Persistence and determination alone are omnipotent. The slogan 'Press On' has solved and always will solve the problems of the human race."

—CALVIN COOLIDGE

CHAPTER 9:

BLOOD, SWEAT, AND ENGINEER$

NO SKILL WAS EVER as important as this last one that I learned in engineering school. Persistence. Engineering school is tough. Period. If you're going through it right now, I feel for you. It does get slightly easier, however, as you delve more into your interests and applicable studies. Persistence, grit, determination, perseverance—whatever you want to call it. This one attribute became paramount.

Before I started school in 2008, my dad and I drove up to Fayetteville for the freshman engineering orientation at the University of Arkansas. This was an exciting time for both of us. My father got to send me off to college, and I was more than happy to go and experience it. When we arrived, they maneuvered us into one of the large auditoriums in the Bell Engineering Center, where most our classes would be held. We found a seat among the hundreds of other eager new students and their parents. After they described the various engineering disciplines and courses, they laid it on us hard and heavy. They described that most of the forthcoming students would not make it out of the university with an engineering degree. This was tough to hear. There was a lot of nervous shifting in chairs and anxious coughs while everyone digested what the dean had just stated. But this was exactly what I needed to hear.

Like most engineering students in college, high school was easy for me. I rarely studied, so I never really knew how to study. I would say that I learned quickly when I made the leap to higher education, though. With classes like Chemistry 2, Calculus 2, and Physics 2, I had no choice but to learn quickly. I began to enjoy the grind associated with getting decent grades. But not everybody had the same affinity for "the grind" as I did.

As I continued through my freshman and sophomore years of engineering school, I began to see what the dean was talking about. Many of the friends I encountered in classes slowly started to drop out. They decided they didn't quite enjoy the pressure of maintaining good grades, while the classes got tougher and tougher, not to mention the fact that a lot of the engineering classes took place early in the morning. I guess they figured engineering students wouldn't be partying as much! I'm not going to say I took the hardest classes on campus, because there were far harder ones in the chemical and biological engineering schools. Most of the students who dropped out did so in the first two years, when everyone in engineering takes the same core classes. After this two-year period, you start to branch off and become more specialized in your degree.

The program constantly tested you. There were late nights studying, missed parties, and missed social opportunities. I did my best to balance my scholastic life with a social life. I joined a fraternity my freshman year; in hindsight, this might have been a bad idea. I did my best to maintain my academic standards while enjoying the social activities involved with the Greek system—and succeeded. I maintained my grades and had a good time doing it. Needless to say, sleep came at a premium freshman year. Throughout all the struggles, including thermodynamics, I came out on top and graduated with my Bachelor of Science in civil engineering. Azim Premji, an Indian business mogul and engineer, once stated, "You have students in America, in Britain, who do not want to be engineers. Perhaps it is the workload; I studied engineering, and I know what a grind it is."

I am grateful for my experience at the university because of the skills I learned along the way. The hard work and determination needed to pass the classes were probably the biggest takeaways from my education. The classes were meant to teach the various aspects of engineering, but they ended up teaching much more than that. Once I graduated from school, this became even more apparent. I would like to repeat the quote I inserted at the start of this chapter because it has meant so much to me. Calvin Coolidge has been attributed to this "persistence" quote but has never actually been confirmed to have written it. Regardless, it resonates with anyone who is working toward a seemingly unobtainable goal: "Nothing in the world can take the place of persistence. Talent will not; nothing is more common than unsuccessful men with talent. Genius will not; unrewarded genius is almost a proverb. Education will not; the world is full of educated derelicts. Persistence and determination alone are omnipotent. The slogan 'Press On' has solved and always will solve the problems of the human race." I challenge you to recall these words when you need them most. I guarantee they will help you through tough times.

According to a study by the Higher Education Research Institute at UCLA, "More than 35 years later, in 2009, approximately the same proportion of students reported intentions to major in STEM on the 2009 CIRP Freshman Survey as in 1971." The proportion was roughly 31 percent of all incoming freshman who declared a STEM (Science, Technology, Engineering, Math) major finished college with a STEM degree. This number is staggering! I think this alludes to a far bigger problem than just the educational system. If you really work for the degree, you can achieve it. It doesn't matter what your background is or what high school you came from. If you talk to any graduate in these fields, I'm sure you will hear the same thing. They all can attribute their success to never giving up on the goal of graduating with this degree.

In my career, I have encountered many adversities and hardships. Many of the projects I have worked on have come with their fair share of difficulties. These situations aren't always found in the designs themselves; they could come from a difficult client, accelerated deadlines, or even unrealistic expectations. The guiding principle of persistence has carried me through these trying times.

When discussing the topic of perseverance with others, I'm often reminded of the TED Talk given by Angela Lee Duckworth, author of the *New York Times* best-seller *Grit: The Power of Passion and Perseverance.* In her famous talk, she discusses the true gauge of success in individuals. She states that "one characteristic emerged as a significant predictor of success, and it wasn't social intelligence, it wasn't good looks or physical health, and it wasn't IQ. It was grit." She goes on to define this term *grit* as the "passion and perseverance for very long-term goals. Grit is living life like it's a marathon, not a sprint."

Unfortunately, in society today, it is acceptable to quit. If times are tough, you will see several peers drop out, just like they did my freshman and sophomore years of college. I would not attribute the success of attaining an engineering degree with just intelligence. Many of my peers who dropped out were extremely intelligent. The dividing attribute was a lack of self-discipline and persistence. It's disheartening to see such bright minds so close to achieving their goal being held back by their own negative self-talk.

You can apply this persistence mantra to yourself and your company. A lot of business managers will shudder at the thought of the Great Recession. This last economic downturn is still fresh in the minds of many hardworking Americans. From 2007 to 2009, the country was rocked by an $8 trillion housing bubble. Many companies could not survive through this tragic economic period. Development ground to a halt while banks were told to freeze lending. The ruins of these bankrupt developments are still visible today. This era didn't just teach us about greed and gluttony, it also taught us about the grit and tenacity of American businesses. Every industry was rocked by the recession. Many businesses were forced to shut their doors. The ones that didn't relied on persistence and ingenuity to ride out the downturn.

So, is persistence just inherent in you or is it a skill you can learn? According to a BYU study, a father's parenting style has much to do with the amount of persistence a child displays. Randal Day, a professor in the School of Family Life at BYU, stated, "Fathers have a direct impact on how children perceive persistence and hope and how they implement that into their lives." The study was focused mainly on the impact of parental mentoring in the lives of preteen and teen children. This does tell us one important truth, though.

Persistence is not related to any certain gene. It appears to be something that is learned. This should give you hope if you are not an overly driven individual. Too often, people blame "bad genes" for deficiencies.

So how do you become persistent? There are many ways to improve your persistence but let me warn you—it's not an easy path to follow. Through my research and personal experience on the topic, I have found a few techniques that seem inherently important when building persistence. These action items include the following:

1. Research the story of hard knocks throughout history.
2. Know your "why."
3. Find accountability.
4. Come to expect "the grind."
5. Have a vision.

This list may surprise or scare you a little, and that's perfectly all right. Throughout engineering school, I used these tactics to help build my propensity for persistence. Much of this is just a mind shift from what you would normally come to expect. Let's delve a little deeper to see how you may be able to incorporate these practices into your daily life to break through tough situations.

Research the Story of Hard Knocks Throughout History

The history books are full of politicians, businesspeople, and religious figures that have experienced hardships much worse than most people can even imagine. When you research these figures, you can see that one principle has helped them succeed—persistence. Persistence enabled Stephen King to succeed after thirty rejections. The story goes that after having his first book rejected thirty times, Stephen King threw the manuscript in the trash out of frustration. His wife then retrieved it and urged him to continue. Stephen King's books have since gone on to sell over 350 million copies.

Walt Disney faced difficulties early on as well. He was fired from his job at the *Kansas City Star,* a local newspaper, because he "lacked imagination and had no good ideas." If that weren't a big enough blow, the first company he started was forced to file for bankruptcy because he lacked the business savvy to maintain multiple salaried cartoonists. As you know, he didn't stop after that failure. He moved on to Hollywood to start from scratch with the knowledge he gained from his previous failures. He would then go on to cofound Walt Disney Productions with his brother. As the famous radio host Paul Harvey would say, "And now you know the rest of the story."

These stories are just the tip of the iceberg. The hard truth about this tip is that everyone who has made history has faced hardships. This will become more evident as you do research for yourself. All of the people hallowed throughout history have encountered resistance. The reason they're in the history books is because they overcame this resistance. **The people who gave up when faced with resistance are the ones who were never remembered.**

Know Your "Why"

Most businesses have a mission statement. Most of these mission statements are filled with grandiose words that are actually very hollow. But there have been a few over the years that are actually inspiring. See if you can recognize any of these well-known companies in the following list. The answers follow on the next page.

1. "All the music you'll ever need is right here. Your favorite artists, albums, and ready-made playlists for any moment."
2. "Help people discover things they love and inspire them to go do these things in real life."
3. "Belong anywhere."
4. "Transportation as reliable as running water, everywhere for everyone."
5. "________ was founded under the belief that a future where humanity is out exploring the stars is fundamentally more exciting than one where we are not. Today, ________ is actively developing the technologies to make this possible, with the ultimate goal of enabling human life on Mars."

The answers to the previous mission statement quiz are as follows:

1. Spotify
2. Pinterest
3. Airbnb
4. Uber
5. SpaceX

So, how'd you do? I'd imagine you got a couple of them right because of their clear statements of mission—their "why." If I were working for one of these companies, I would know what the ultimate goal of the organization was.

Your goal is to come up with a mission statement for your personal and professional life. By understanding your true mission, you will more likely be willing to persevere through various obstacles to fulfill your mission. This mission statement must be bigger than your emotions or feelings.

In Steve Jobs's famous commencement speech at Stanford University in 2005, he stated, "You've got to find what you love. The only way to do great work is to love what you do. If you haven't found it yet, keep looking. Don't settle. As with all matters of the heart, you'll know when you find it." He states that the only way to do what you love is to find it. Is that how it really works? Do you have to search for your passion or meanings, or is it something you run into like a brick wall?

I can see both sides. Growing up, I never quite knew what I wanted to be—I didn't even know what I was truly passionate about. In school, it was genuinely difficult for me to say that I was really passionate about anything (other than myself, I guess). I would spin this to say that you can become passionate about your mission in life, not necessarily a career. But once you understand how you can align your career with your mission, then you will truly thrive. Once I made the mental shift to change my career into more than just a paycheck and into a vessel that would align with my personal mission, I really began to love what I was doing. It seemed to have purpose. *From purpose came passion.*

Find Accountability

When attempting to lose weight, many adults turn to accountability relationships for an extra push. In a study published in the *Journal of Consulting and Clinical Psychology*, a group of 166 adults were enrolled in a weight-loss program. One group was enrolled with friends and given additional social support. Not surprisingly, 95 percent of this group completed the program and 66 percent of them maintained the weight loss for a period of at least six months after the program was finished. The group that completed the program without the social accountability had a success rate of 76 percent, and only 24 percent maintained this loss at the six-month post-completion mark.

There are many more studies that show the same results. The impact of accountability can be seen just about anywhere. On social media, many people post non-flattering "before" pictures when starting a diet. Alcoholics Anonymous (AA) uses social accountability to help its members persist through alcohol cessation. I used this tactic when writing this book. I told several of my peers and family that I was going to write a book. This held me accountable to my actions because I didn't want to be wrong or lie to my family and friends. Accountability can be a very important tool when you find that internal discipline is not enough to persist through certain situations.

Come to Expect the Grind

No man or woman has faced a grind as daunting as putting together IKEA furniture. You see it online or in the store and think to yourself, "How bad could it be?" Famous last words. When you have the four-by-eight-foot box sitting in your living room, it quickly becomes hard to see how this will become a bookcase. Expectation is a cruel mistress. Once you open the box, you find your expectation of this process doesn't match reality. Six hours later, you have a piece that looks "close enough" to the picture on the box.

This is a trivial example, but it's one that most all of us can relate to. We expect tasks to take certain amounts of time based on previous experiences or recommendations. These expectations set us up for victory or failure. When you assume things are going to be easy, you set yourself up for the possibility of failure.

Have a Vision

This heading may seem similar to a previous point, "know your why," but it's actually very different. Having the vision to pursue your dreams is more about seeing into the future than understanding your purpose. It's about taking your mission statement and visualizing what that could look like in the future if you follow it. If your own vision doesn't reflect your goals or mission, it's unlikely you will make them a reality. You have to believe you can make it happen. This sounds cheesy, but let me give you an example or two.

I often ride bicycles in my free time. It's one of the ways I love to keep in shape. One day while I was riding along a trail that looped around a nearby lake, I noticed a bollard in the middle of one of the pedestrian bridges. A bollard is one of those reinforced posts you see near buildings or other structures to ensure that vehicles do not interfere with them. I understood that this bollard was most likely to keep cars from crossing the bridge, but for some reason I became fixated on it. As I was coming down the hill toward the bridge, I couldn't stop focusing on this unwavering yellow bollard in the middle of the trail. As a result, my bike started to veer ever so slightly toward it. As I careened toward it, I noticed that my trajectory was headed straight for the bollard. I came to my senses at the last moment, jerked the handlebars away from that path, and narrowly missed the bollard. You don't have to ride a bike to understand this phenomenon. I have noticed the same thing while driving on the interstate. I'm sure you have, too. When I first started driving, while passing semitrucks on the interstate, I would get fixated on the large tractor trailer beside me. My car would veer ever so slightly toward the truck at 75 mph. Then I would wake up from my highway-induced trance and jerk the wheel away from the slower tractor trailer beside me.

These real-life examples help paint the picture of what fixation and vision produce for you. When you keep your eyes locked on a certain object, you often seem to gravitate toward it. Just like myself on the bicycle heading toward the bollard or in the car veering toward the semi, we often get fixated on things that will harm us rather than the things that we truly want.

How long has it been since you daydreamed about your future? What was that like? What did you dream about? My guess is that it's been awhile since you have truly imagined what your future could be like. Dreaming about your future not only excites you and gives you hope when facing tough times but also creates a reality for you that seems obtainable. You will become fixated on this reality, and as a result, opportunities that you once discarded will appear as stepping-stones to the place you want to be.

What if everyone tried just a little bit harder? Persisted a little longer? What if you did the same? The limits of what the human race could do would be endless. Your limits would be endless. What if you got into the minds of all of those who have persevered through hard times? What if you came up with a mission statement that propelled you to heights once thought unimaginable? What if you created an accountability group of like-minded individuals and learned to love the grind? There are a lot of "what ifs" presented here, but every one of them is doable. You just have to change your mindset. It's like Mark Cuban once said: "It's not about money or connections. It's the willingness to outwork and outlearn everyone when it comes to your business."

CHAPTER 9 SUMMARY

BLOOD, SWEET, AND ENGINEERS

People love hearing success stories. We love to hear how these rare individuals beat the odds. But how can we create that same kind of success? Build your persistence with the following techniques:

1. **Research the story of hard knocks throughout history and understand that your trials often don't compare to those in the history books.**
2. **Know your "why." Have a mission statement bigger than yourself.**
3. **Find accountability in social groups that provide support for your endeavors.**
4. **Come to expect the grind associated with achieving your dreams.**

Now we'll take a step outside the classroom setting and delve into the skills learned in the real world.

"The competitor to be feared is one who never bothers about you at all but goes on making his own business better all the time."

—HENRY FORD

SECTION 3:

Business Skills Learned Outside the Classroom

“Management is doing things right; leadership is doing the right thing.”

—PETER F. DRUCKER

CHAPTER 10:

SUN TZU AND COMMANDING THE APOLLO 13: THE ART OF PURE LEADERSHIP

On a cool afternoon at the Kennedy Space Center in Florida, the crew of the Apollo 13 was preparing for liftoff. The crew, consisting of James A. Lovell Jr., John L. Swigert Jr., and Fred W. Haise Jr., were getting ready for launch. At mission control in Houston, the usual flight checks were being completed. At 2:13 p.m. eastern standard time, the Saturn V rocket propelled the astronauts up to the heavens at 17,500 mph.

This flight was unlike the previous two lunar flights. Apollo 11 and 12 had both landed on the moon successfully the year before. Unbeknownst to the crew and mission control, this would become a story of extreme leadership, creative engineering, and unparalleled resourcefulness.

The mission was going almost better than expected the first few days following liftoff. The crew had just filmed a television broadcast when an explosion rocked the command module. No one was sure what might have happened. Warning lights started flashing all over the command module. They would find out later that oxygen tank 2 had exploded and oxygen tank 1 had malfunctioned due to the explosion. This was roughly fifty-six hours into the mission and approximately two hundred thousand miles from Earth. As Commander Lovell peered through the window, he saw something shocking. The remaining oxygen tank was leaking. He radioed mission control, "We are venting something out into the . . . into space."

The explosion had damaged two of the three main fuel supplies. These fuel supplies helped to feed the normal source of electricity, water, and air to the module. With the oxygen constantly draining, the crew knew they wouldn't be landing on the moon this time. Their only concern was making it back to Earth.

Mission control became frantic as the catastrophe unraveled in front of them. Flight Director Gene Kranz now had the task of leading the talent of NASA to bring these men back to Earth. Kranz was a pilot and engineer himself. At thirty-one, he was one of the oldest flight directors employed by NASA. He first calmed the men at mission control, never letting them believe that the astronauts were a lost cause. He believed there was a solution and exuded this confidence to those around him. He then gathered all the data he could from the ship's computers. After they had synthesized and prioritized the problems, he delegated the work of coming up with solutions to the most capable engineers. These engineers then followed the problem-solving steps I mentioned previously in this book to come up with solutions to save the astronauts.

Commander Lovell suggested that the crew board the connected lunar module, which was still fully intact. This would become their proverbial lifeboat. The lunar module had enough oxygen to last the entire trip back. Power was a concern, though. They were able to charge the batteries on the lunar module from the remaining power on the command module. They were also able to reduce the power to one-fifth of what was previously being consumed. The next problem they faced was a water shortage—they wouldn't have enough water to cool the systems at the rate they were consuming. They began to cut the water back to one-fifth of what was normally consumed. They relied on wet-packed foods and juices to help provide additional hydration.

Now that they had enough oxygen, they had to determine the best way to remove the excess carbon dioxide (CO_2). The lunar module was equipped for two passengers, not three. The three crewmen produced more CO_2 than the lunar module could process. The process NASA relied on included the use of lithium hydroxide (LiOH). LiOH, when combined with CO_2, creates water (H_2O) and nontoxic lithium carbonate (Li_2CO_3). The command module had sufficient LiOH canisters for the crew, but these canisters were square. The lunar module used round canisters. As some would say, it's hard to fit a square peg in a round hole. The engineers at NASA realized they needed a solution fast so the astronauts wouldn't be subjected to a fatal concentration of carbon dioxide in their cabin. Kranz gathered the engineers and explained the problem. He then let them use their creativity and anything they would normally find onboard the lunar module to create a device that would solve their square-peg-round-hole issue. The device they engineered was a mix of duct tape, tubing, cardboard, and plastic. They created a prototype of the makeshift air scrubber and tested it. After sufficient testing, they sent the directions to the Apollo 13 crew. The crew was able to recreate the apparatus, which ultimately bought them more time in the lunar module. But there was still one last hurdle—the men needed to return to Earth safely.

They had to position the spacecraft back into the moon's orbit to "slingshot" the craft around the moon to get back on their free-return course to Earth. The crew would be traveling on the dark side of the moon with little to help them navigate. The ground crew quickly formulated a plan. They were able to use monuments on the moon's surface to help navigate. Ultimately, the crew was able to return back to Earth unharmed.

The story of Apollo 13 is one of courage, tenacity, and ingenuity. With the leadership of Gene Kranz, the ground crew was able to efficiently solve problems given the constraints imposed on the crew in space. Kranz never doubted that he could bring the crew back safely. He understood that he didn't have all the answers needed to do this by himself. He also knew it was his responsibility to fix this mistake. He understood that even though it might not have been his fault that the ship failed to make it to the moon, he was still accountable for the outcome. Kranz never allowed the stress of the situation to show when addressing his subordinates. His demeanor went a long way toward assuring the ground crew that they would find solutions to solve the task at hand. In the following chapter, we will analyze these leadership techniques employed by Kranz, as well as a few additional practices I have found important and effective when leading a team, no matter the size.

Be Confident

Kranz never once doubted the success of the mission at hand and neither should you. As a leader, your emotions are often visible to your subordinates. If you believe a mission might be impossible, it will show in your attitude and emotions. This mindset will be contagious to the rest of your team.

I've found this trait difficult to adopt, being so green in my field. I would get questioned on a certain design element that I knew was right, but when interrogated, I would become unsure of myself. It's difficult when you're one of the youngest team members, but it's necessary to be sure of your ideas. When you're new to an industry and called on to be a leader, I would recommend that you don't fake confidence in your skill set. People can see right through this deception. Do your best to ask the right questions and gain the knowledge necessary to fulfill your position. Learn as much as you can so you can be confident in your decisions. Decisiveness is a clear identifier of confidence.

> ***"If you can't stand the heat, you'd better get out of the kitchen."***
>
> ***—HARRY S. TRUMAN***

Be Accountable

One of the worst things you can do as a leader is to push accountability onto one of your subordinates. Unfortunately, I've seen this happen many times in my career. You are ultimately responsible for the outcome of the team. If the team causes any errors, the team leader must take responsibility. This will create a culture of accountability in your group. If you, as the team leader, are accountable for the mistakes of a group, the individuals in the group will become more accountable for their work. Ultimately, this is how a successful team functions.

Be Motivating

In my various leadership positions, I have found this to be extremely important. If your subordinates don't appreciate—much less understand—the vision of the team, they will not flourish. I've written bylaws for several nonprofit groups over the years. The vision statement often gets glossed over when others read the bylaws. It's important that leadership understands the goals of the organization and transmits them to the members of the group. This mission should motivate and invigorate the group members. It creates a commonality among everyone involved.

Lead by Example

As a leader, it is easy to fall into the trap of leading by command rather than by example. Once you have attained the role of leader, whether by appointment or natural progression, it becomes easier to just shout orders than to truly understand the work involved in your position.

A lot of this comes from empathy. Once you empathize with your subordinates, they will be much more willing to follow your instructions. You can gain respect through empathy. When you take the time to understand the task you are asking of subordinates, you will often gain their esteem.

One of my favorite stories about leading by example goes like this: On a battlefield nearly two hundred forty years ago, a man dressed in civilian garb was riding his horse down a path. The man passed a group of soldiers tiredly digging a pit. The soldiers were worn down by the physical labor, and their uniforms were covered in mud and soot. A man who appeared to be their commanding officer was sitting atop a horse nearby shouting orders, obscenely barking threats of punishment if this entrenchment was not dug prior to the ensuing attack. As the stranger on the horse started to pass the officer, he asked, "Why are you not helping these poor soldiers?" The officer barked back, "I am the officer; these men will do as I tell them." He then asked the stranger, "Why don't you help them if you feel so strongly about this?" So the stranger got off his horse and joined the men in the trench. He continued to help the weary men until the task was finished. Before leaving, he congratulated the men and praised them for their effort. He then mounted his horse, approached the confused officer, and said, "You should notify top command next time your rank prevents you from supporting your men, and I will provide a more permanent solution." As the stranger turned to leave, the officer noticed who it was. The commander in chief, George Washington, had just given this rash officer a lesson in leadership.

Many people have likened business to war over the years because of its need for strong leadership figures and strategy. I can agree with some of the similarities proposed. If you're interested in these similarities, you should pick up *The Art of War* by Sun Tzu. Often touted as a business strategy book in disguise, *The Art of War* has several lessons that can be utilized when "controlling your troops" in the office. A couple of my favorite lessons from the book on leadership include: "Regard your soldiers as your children, and they will follow you into the deepest valleys; look upon them as your own beloved sons, and they will stand by you even unto death. If, however, you are indulgent but unable to make your authority felt; kind-hearted but unable to enforce your commands; and incapable, moreover, of quelling disorder, then your soldiers must be likened to spoilt children; they are useless for any practical purpose." These are very powerful messages.

When you treat your subordinates as family members by showing compassion and empathy, you can gain their trust. If they trust you, they will follow your command. This trust is not won easily, but once obtained, it's powerful when tackling problems in the workplace. There are many more valuable lessons in that short, historical guide by Sun Tzu, and I recommend delving into it with an open mind. The book is often touted for its strategic approaches in war and business, but the other important lesson in this book is on strong leadership.

We've seen how many great leaders throughout history have handled leadership. What if you utilized some of these techniques in your day-to-day leadership? What if you implemented some of the same tactics as our first example, Flight Director Gene Kranz? You will need to come up with your own style of leadership, but remember to build it on a foundation of respect. Next time your team shouts, "Houston, we have a problem," remember the lessons learned from leaders throughout history and be ready to tackle any problem that gets thrown your way.

CHAPTER 10 SUMMARY

SUN TZU AND COMMANDING THE APOLLO 13: THE ART OF PURE LEADERSHIP

Problems arise every day. It takes strong leadership to position the right people to take the right action at the right time. Just as Gene Kranz navigated that interstellar conundrum, we must step into this position and manage our teams. To be effective as a leader, it's essential to:

1. Be confident you will succeed.
2. Be accountable for your team's failures.
3. Be motivating to your team members.
4. Lead by example.

By utilizing these techniques, you can safely bring your team back from any problems you encounter. Leadership can often cause anxiety for some people when the team is on the line, but in our next chapter, we will delve into a topic that is even more frightening for most.

"There are only two types of speakers in the world.

1. The nervous and
2. Liars."

—MARK TWAIN

CHAPTER 11:

THOU SHALT NOT SIMPLY TROT OUT THY USUAL SHTICK

TOMORROW, YOU ARE TO give a speech on the history of the Franco-Prussian War. You will be giving this speech in front of your peers and superiors in the boardroom. Your speech will need to be at least thirty minutes long and cover the war and its impact on the Western world.

What are you feeling right now? Is your mouth dry? Are you shaking? Is your heart rate increasing? Sweating a little bit? Stiffness in your neck and upper back? These are all symptoms of glossophobia, the fear of public speaking. If you're like most people, you're deathly afraid of public speaking. Studies show people are more afraid of public speaking than any other thing—even death. So, how do we get over this fear? There are many steps that can alleviate this stage fright, but I'll get into those later in this chapter.

The title of this chapter comes from the TED Commandments. The TED Commandments are sent to all upcoming speakers prior to giving a speech at the annual TED Talk events. The speeches and the commandments highlight the importance of a well-presented speech in shaping the world. You may not be giving a TED Talk anytime soon, but the impact a superbly given speech can have on your career should not be downplayed.

Why is it important for you to learn how to speak publicly? As a business-person, you may need to ask for funding or need to address a group of investors. You might be put on a panel of experts, or you may be asked to explain your business goals in a television interview. As an engineer, we also need to know how to give speeches. Engineering speeches are more related to sharing technical ideas with others. As an engineering consultant, we often give speeches about a client's project or can be called to testify as expert witnesses.

Public speaking is a very important skill to master. It takes time, it takes effort, and it often takes many subpar speeches on the way to becoming a successful presenter. I can remember one such experience when I was enrolled in an entrepreneurship class in college. For some reason, I—the engineering major—was "volunteered" to be the presenter for our group's class project. With a group of finance and marketing majors, I was asked to give the elevator pitch for the group. They obviously weren't aware of the lack of experience undergraduate engineers have in public speaking.

I was terrified. When I got up to speak, my palms started to sweat, and my breathing became short. I told myself it was "game time" to psych myself up. Real "Eye of the Tiger" type stuff. I presented our project with as much enthusiasm as I could muster. I was not completely sold on the business idea we were pitching, but I wanted to make sure we came off as credible. I stumbled through the pitch with as much tenacity as I could muster. Looking back, it was an absolute train wreck. We took questions at the end, which were more like personal shots at our idea than actual questions related to our entrepreneurial endeavor. After the dust had settled, there was little left of our fragile egos. I'm being very dramatic here, but it was not a good presentation by any means. Needless to say, we did not win the competition, but it was great experience for my future public speeches.

Unfortunately for engineering students, college does little to prepare engineers for the role of public speaker. Engineering schools teach the fundamentals and technical knowledge to become engineers, but they're not required to give public speaking classes, and, unfortunately, often can't fit them into the curriculum. As an engineer, I joined the business college to obtain a minor in business. Public speaking was a prerequisite for this minor. That turned out to be the only public speaking class I had to take at the time. It also turned out to be one of the most valuable classes I took. So how do you give this perfect speech that is so integral to your career, not only as a businessperson but also as an engineer?

Pre-speech Preparation

Before you even consider giving a speech for your company, I recommend joining a public speaking club like Toastmasters, a US-headquartered nonprofit educational organization that operates clubs worldwide for the purpose of promoting communication and public speaking. By getting involved in these types of clubs, you can start gaining experience with low-stakes speaking opportunities. Public speaking is a skill that requires a lot of practice for most people. By joining these clubs, you can practice for these nonessential speeches that won't affect your career. It's a great starting point for your future speeches.

Dale Carnegie stated this best in his book *The Art of Public Speaking*. When asked about how best to prepare for public speeches, he gave this response: "Students of public speaking continually ask, 'How can I overcome self-consciousness and the fear that paralyzes me before an audience?' Did you ever notice when looking out a train window that some horses feed near the track and never even pause to look up at the thundering cars, while just ahead at the next railroad crossing, a farmer's wife will be nervously trying to quiet her scared horse as the train goes by? How would you cure a horse that is afraid of cars—graze him in a backwoods lot where he would never see steam engines or automobiles, or pasture him where he would frequently see the machines? Apply horse sense to ridding yourself of self-consciousness and fear: Face an audience as frequently as you can, and you will soon stop shying. You can never attain freedom from stage fright by reading a treatise. A book may give you excellent suggestions on how best to conduct yourself in the water, but sooner or later, you must get wet, perhaps even strangle and be 'half-scared to death.' There are a great many 'wetless' bathing suits worn at the seashore, but no one ever learns to swim in them. To plunge is the only way." In short, he simply states that there's no better substitute for practicing public speaking than public speaking itself. The best public speakers are the best because they have failed more often than you have. Keep that in mind when you're on a mission to becoming an excellent public speaker.

Once you get a couple of speeches under your belt and you come up to your next speech, you want to prepare and practice your topics. You want to know your material thoroughly; you want to know when to inflect your voice and when to give certain signs or cues. Knowing your material completely will help your confidence. Don't memorize your material; just know it. It's extremely uncomfortable for your audience to watch you read your script on stage. It shows a lack of knowledge and credibility. People who look calm and relaxed on stage are confident because they know everything about their topic. The passionate speaker is the most interesting speaker. When you know the material and you're passionate about the topic, the audience will reciprocate your enthusiasm.

Now that you're prepared and you know all your material, it's time for the big day. During a speech, there are so many things you can do to remain calm. One of them is to simply practice breathing—a slow, deep, downward breath—before and during the presentation. Basically, relax your body.

Game Time – Giving the Speech

Now, during the speech, you don't have to block out the audience or imagine them in their underwear. Try viewing them as supportive individuals trying to help you give your speech. I think once you can get this paradigm shift in your head, you begin to see them as cohorts and not as people trying to cast you out.

In all the research I can find on this topic, the one tip that connected all of the greatest public speakers, such as Simon Sinek and Tony Robbins, was that they actually wanted to connect with the audience. This connection was made by maintaining eye contact with various members of the audience, as well as truly understanding what they could gain from the speech. In an interview with *Entrepreneur* magazine, Sinek stated, "It's like you're having a conversation with your audience. You're not speaking *at* them, you're speaking *with* them." This is a different way of thinking for most speakers, myself included. I often see the audience as a foe I need to sway toward my ideas. This is the exact opposite method utilized by some of the most successful speakers.

Nonverbal cues are everything in a speech—the way your face appears, how your body language conveys your message—these are very important to how your audience will receive the message. There are many books and videos on the proper way to conduct your body in order to convey a certain message. Nonverbal cues are essential; these can include pauses, voice inflection, and physical gestures. All of these things give your audience another way to interpret your message. You want to make sure your verbal message matches that of your nonverbal message. If they're incongruent, people may distrust your message. For example, if you're relaying a somber message on the death of your pet to someone and you're smiling, the person has obvious reasons to doubt your story because your verbal and nonverbal messages don't match. If you're giving a speech on a field of study that you are a leading expert on, but you're slouching or you have your hands in your pockets or show other signs of lack of confidence, the audience may doubt your expertise.

As a speaker, it is your duty to convey a message that benefits the audience, be it a wedding toast, award presentation, or technical seminar. Tony Robbins once told interviewers with *Business Insider* magazine, "Don't ever speak publicly about anything that you're not passionate about and that you don't actually believe you have something truly unique to deliver. Don't get roped into talking about something that you don't really have passion for, and don't get roped into something you don't have expertise in. Why should somebody listen to you? If you're going to take somebody's time, you better deliver." Once you see your speech as a chance to deliver valuable information to the audience and you shift your mindset outwardly, the speech becomes easier. *You're no longer the focus of the speech; the message becomes the focus.* This takes a lot of pressure off you, the speaker.

If you're uncertain on how best to deliver a message to an audience, take some time to watch YouTube videos of certain outstanding presenters, such as Martin Luther King Jr., John F. Kennedy, Tony Robbins, Steve Jobs, Eric Thomas, and Jim Rohn. Watch their physical, nonverbal communication. Pay attention to their delivery, pauses, and vocal inflections. Pick one of these speakers as your "speech mentor." Try your best to emulate your mentor and practice their movements, gestures, and other nonverbal cues.

This chapter is not meant to be an all-inclusive dive into the critiques of proper speech presentations, but it is meant to shift your perspective on certain aspects of giving speeches that should make your presentation overall more successful.

CHAPTER 11 SUMMARY

THOU SHALT NOT SIMPLY TROT OUT THY USUAL SHTICK

No one ever made it to the top by being comfortable. You will have to step out of your comfort zone and give a public speech during the course of your career. Why don't you take the time to prepare for it now? Below are some tips in preparing for your next big speech:

1. Start out by giving low-stakes speeches at your local Toastmasters or another organization you are active in.
2. Know your topic; become an expert in your field.
3. Become passionate about your topic. Feel like you have something valuable to give the audience.
4. Study the great public speakers of our time, such as John F. Kennedy, Martin Luther King Jr., and Tony Robbins.

To truly get ahead in your career, being able to articulate and transmit your ideas to the masses will be very valuable. In the next chapter, we will discuss how you can further transmit yourself to the people who may not know you yet.

"Brand is just a perception, and perception will match reality over time. Sometimes it will be ahead, other times it will be behind. But brand is simply a collective impression some have about a product."

—ELON MUSK

CHAPTER 12:

THE NIKE SWOOSH AND MARKETING YOUR OWN BRAND

IN 1997, THE FOUNDERS of Google, Larry Page and Sergey Brin, coined the name of this iconic company. The name Google came from a play on the word "googol," a mathematical term for the numeral 1 followed by 100 zeros. According to Google's website, the founders picked this name because it "reflects Larry and Sergey's mission to organize a seemingly infinite amount of information on the web." We immediately recognize the iconic brand and what it stands for. We trust that when we type a question or query into that search bar, Google will give us a seemingly endless number of relevant results.

Or, how about the famous shoe brand that has helped propel famous athletes all over the world—Nike, named after the Greek goddess of victory. Nike's iconic "swoosh" was designed before the name of the company was even determined. The founder of Nike, Phil Knight, had enlisted the help of Portland State University graphic design student Carolyn Davidson to come up with some design options. One of the choices she presented to Knight was the now-iconic swoosh. This logo is universal. No matter where you live, you know that the Nike swoosh stands for quality athletic shoes and apparel. When people talk about you, do they know what you stand for?

In business school, we are taught the five P's of marketing. The five P's are the five attributes of a marketing campaign that can be altered to make a product launch or marketing campaign successful. They are: **P**roduct, **P**rice, **P**romotion, **P**lace, and **P**eople. These attributes can be used to describe your *market share* in a given space as well. Your *product* is the knowledge and skill sets you bring to the arena. This can constantly be improved through lifelong learning. Your *price* is how much value you bring to your organization. Your *promotion* is your ad campaign. What are you doing to broadcast your product or your skill sets? Your *place* is commonly referred to as the channels in which you utilize your product. Do only the people in a small town know your accomplishments and understand your talent, or are they known throughout the world? Your *people* refers to the human element that is inherent in the market. Do you know and understand people? Are you respected in your market? These five P's can change the way you look at how you market your own skill sets. With engineers, the hardest part of this marketing equation is promotion.

Many engineers do not shamelessly promote themselves. As someone who never promoted my own accomplishments before, I can relate. I always thought people who "tooted their own horn" were arrogant and irrational. However, this is your career, and many of your colleagues won't take note of your accomplishments unless you broadcast them. I'm not saying that you should constantly remind your peers of the numerous accomplishments you have achieved; I am simply saying you should be proud of your accomplishments and not be bashful of letting others know what you're capable of. As with accomplishments, your values should also be visible to others. Your values will be challenged often in your career. It's essential to be able to stick to your values to gain respect from your peers. This is extremely important not only in engineering but also in business. With the advent of social networking, marketing your own brand and values has never been easier. In the following pages, we will discuss ways that you can build your brand so that others will identify you immediately. Similar to the colorful Google icon, your brand will be sought out for answers if you follow the next steps. See the following flowchart on the natural progression for building your own personal and professional brand. This is not set in stone; these processes can be done simultaneously and should be—doing so will help your marketing reach multiply.

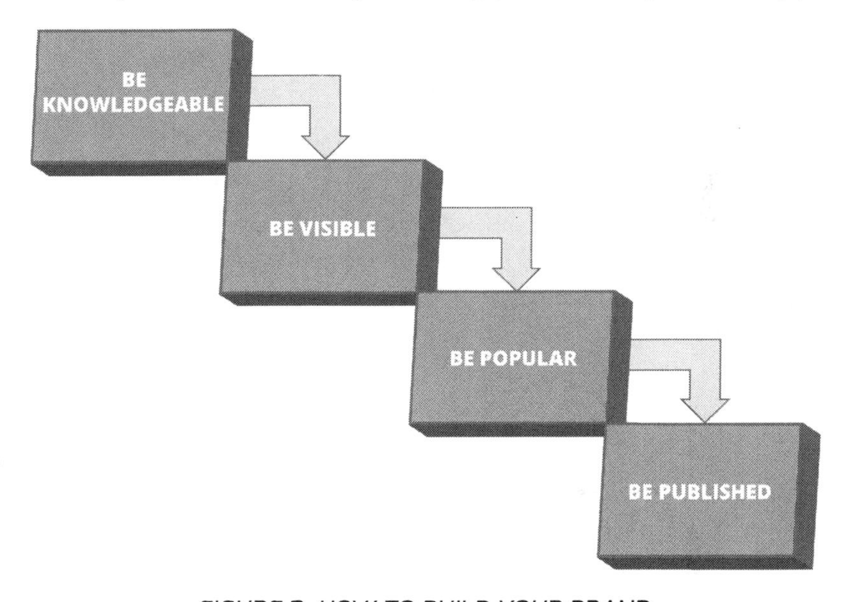

FIGURE 2. HOW TO BUILD YOUR BRAND

Be Knowledgeable

Become a leader in your field of choice. Never quit learning skills and technologies that add value to the marketplace. People seek out knowledgeable professionals when looking to solve their problems. If you can demonstrate your aptitude in a certain field, especially engineering, you will be sought out by others looking for your expertise. To obtain this technological knowledge, subscribe to trade publications, take professional development classes, or even go back to school, if deemed necessary. This is a beautiful world we live in where we can access information, often for free, almost instantly and anywhere in the world. There really are no excuses for failing to obtain technological knowledge.

Being knowledgeable doesn't just touch on technical know-how. It can be an increased knowledge and presence of a certain area or industry. Let's say you're knowledgeable in a certain city regarding its movers and shakers in government. You may be sought out to obtain approval on a project or request due to your expertise. To be knowledgeable in a certain area is simple. You just need to make a concerted effort to get out in the municipality or group. Expose yourself to society and make yourself visible. Become vulnerable to these experiences and there's no doubt you will obtain invaluable contacts along the way.

Be Visible

This one may be all too obvious, but it's extremely relevant for most engineers. Most engineers I've met appear to enjoy staying out of the limelight. Most enjoy the simple pleasures of solving problems on a computer or behind a desk. If you're not visible to others, they obviously won't notice you. Some ways I achieve this visibility is through social media, networking events, and attending technical conferences.

Social media often gets a bad rap. People use it to make obscene personal declarations, post cat videos, and post pictures of their spring break exploits in Cabo. So of course many professionals dismiss it. But sites like LinkedIn and Shapr are helping to promote professional networking. I use LinkedIn religiously to understand industry trends, network with similar professionals, and reach out to professionals in other industries that I might not have been able to previously. After hearing a speech from a local business mogul, I reached out to him on LinkedIn. He was gracious enough to meet me in person to discuss various aspects of his business, as well as my own personal aspirations. This contact has become a professional mentor of mine, and I may have never come into contact with him without social media.

With the use of LinkedIn, I also make myself professionally transparent. I broadcast my achievements, as well as current projects. This tool can be used to show potential clients, potential employers, and other industry professionals what you're doing.

Be Popular

By being popular, I'm not insisting that you try to reenact the main characters of the hit movie *Grease*. Being popular goes hand in hand with being visible. Being popular goes much deeper, though. It's not enough to just show up to events; you need to make connections with people within the community or society. By forming these connections, you will be able to broadcast your skill sets much more easily. Your brand will be fortified by firsthand encounters. People will seek your knowledge based on the recommendations given by the network you have created. Building a robust network is what being popular is all about. A network spreads the gospel of you.

Be Published

This one may not seem as obvious but being published or quoted is a great way to expand your personal brand. This is not limited to books or journals; it can mean giving lectures. Even with the use of social media, authorities in certain fields can make their knowledge more accessible to the masses. Whatever platform you decide to use, I recommend you pursue it—be it books, publications, social media, or even lectures. All of these outlets will get you noticed and will help to make you a leader in your desired field.

Think about that book you've always wanted to write. How great would it be to have that book bring you clients? If you're not necessarily in a field where you would have clients, think about the possibility it can bring when you're pursuing a new career path in the future. What if you could put "published author" on your resume? I'm not saying it's easy (as I'm discovering right now), but it's a great way to become more visible.

Have you ever been able to give a technical talk to like-minded professionals? This is another great way to become more visible as a knowledgeable professional. In the last chapter, we discussed the skills needed to master public speaking. What better way to implement these skills than to discuss a topic you know well to a crowd of people willing to listen and learn more?

As you can see, there are many ways to create a brand for yourself to help elevate your career. What if you just implemented half the ideas listed here for creating a brand? I would imagine more and more people would start taking notice.

CHAPTER 12 SUMMARY

THE NIKE SWOOSH AND MARKETING YOUR OWN BRAND

Just as the founders of Google and Nike developed their now universally recognized brands, so must you. Reputation in business and engineering is extremely important. Follow these steps to market the brand of *you*:

1. Be Knowledgeable
2. Be Visible
3. Be Popular
4. Be Published

Marketing your brand is extremely important. But so is getting to know and understand the people who will become your disciples. These people will help spread your message. In the following chapter, we will discuss how to build these lasting relationships that are so integral to success.

"The way we communicate with others and with ourselves ultimately determines the quality of our lives."

—TONY ROBBINS

CHAPTER 13:

IQ OR EQ? NETWORKING WITH DILBERT

WE STRIVE TO KNOW everything we can about our field in order to gain respect from our peers. The relationships I have formed over the years have opened the doors to various opportunities that I couldn't have ever imagined possible. This chapter may take many of you by surprise. Although the thought of forming relationships may make you cringe, relationships are invaluable to your career.

Finally, we come to what I have coined "the Dilbert Effect." Hopefully, you are familiar with the comic strip *Dilbert*, but if you are not, let me describe it to you. This comic strip first came out in 1989 and is written and illustrated by Scott Adams. The comic depicts a typical white-collar work environment that makes jabs at office politics. The main character is an engineer named Dilbert. Dilbert is a quintessential engineer character. He has questionable style, awkward social skills, and is terrible with women. The comic strip depicts Dilbert in all sorts of events in his life, such as getting turned down for dates, getting passed up for promotions, and being disrespected by his boss and upper management.

"The Dilbert Effect" is no shot at the comic strip. I believe that the strip is hilarious and points out many of the flaws of present workplace practices and politics. "The Dilbert Effect" goes much deeper than this, though. I believe that the engineering stereotype has hindered many engineers from truly harnessing their business potential. The archetypical engineer is represented in pop culture as awkward, socially incompetent, introverted, and, if that weren't enough, most of them are shown to be terrible at attracting mates.

Sure, there have been some attempts to reverse this trend with characters such as the insanely rich playboy engineer Tony Stark, better known as Iron Man. Sure, this helps a bit, but if you still ask most people what they picture when you mention engineers, or really anyone in a STEM field, you will get similar Dilbert-like descriptions. Mainstream media over the last few decades has run its course in the hearts and minds of kids and adults alike. The result is a misunderstood and misguided view of the work performed by engineers.

I believe that a majority of engineers have embraced this socially awkward stereotype. From firsthand experience, I have seen how many engineers feel the need to enjoy other "nerd-like" hobbies in order to "fit in" with other engineers. I believe that some engineers even feel that if they don't go along with the stereotypes, they are less of an engineer and somehow not as smart or intelligent as their "nerdy" counterparts. This is flawed thinking, and it ultimately hurts the field as a whole. The need to fulfill this stereotype leads to poor public speaking efforts, lack of social networking, and a holistic deficiency of soft skills that make businesspeople successful.

So how do we encourage more engineers to shake off this stigma related to engineering? This is no easy task, but I believe it is necessary for the profession and society as a whole. If engineers do not become effective communicators and leaders in society, ultimately society will suffer. I believe the best way to make this happen is to encourage engineers into taking leadership positions early in their careers. We also need to shift focus more into the soft skills that rarely get taught in engineering schools these days. By doing these things, I trust we will be able to shift the natural stereotype in the right direction.

By making leadership and communication roles a prerequisite for young engineers entering the workforce, an expectation of these skills will be more prominent. Unfortunately, these skills are found few and far between in engineering schools at the moment. Businesses that utilize engineers will find that if they cultivate these skills early in their developing engineering workforce, the investment will pay off huge dividends. This can be achieved by creating speaking events that introduce engineers to a nonthreatening atmosphere to help get them acclimated to public speaking. This could be further achieved through the expectation of their young engineers to serve on municipal boards or committees. The possibilities for creating such learning experiences are truly endless.

The following pages may be tough to stomach at first for those who are more introverted but keep reading. What many engineers don't realize is that you can't get to the top on your own. This may be a hard truth for most engineers, but it's true. Your knowledge alone can't propel you to the top. Now, I understand that not all engineers strive to be upper management or start their own business, but I'm guessing that if you're reading this book, being a design engineer is not your end goal.

I know I'm making some big claims here, but relationships bring advantages to those who seek them. Contacts bring about other contacts. Your network will constantly be growing. This advantage might not be apparent at first glance. The more people you can reach out to, the better. This network can help you get a new job, bring in new work, answer your questions, and even help spread your "gospel." The key is to form relationships that are mutually beneficial. It is the "you scratch my back, I'll scratch yours" mentality. This is extremely important when forming business relationships. This is the key advantage to creating a large network. When you need help, you have many different resources to pull from.

If you're interested in this topic, I would highly recommend reading *How to Win Friends and Influence People* by the proclaimed expert on interpersonal relationships Dale Carnegie. The passage that stands out the most to me and is most relevant to this topic is based on a study done by the Carnegie Foundation for the Advancement of Teaching. The study states that "even in such technical lines as engineering, about 15 percent of one's financial success is due to one's technical knowledge and about 85 percent is due to skill in human engineering—to personality and the ability to lead people." This finding only reinforces what I had suspected all along. This fact is often the reason many engineers can't get ahead of their more relationship-savvy colleagues. So, take a note from Mr. Carnegie and take a step toward understanding the skills needed for "human" engineering.

So, how do you form these relationships? If you're not comfortable with building relationships with others, I can relate. My story is probably not different from your own. I've always considered myself an introvert. I don't typically go out of my way to meet new people. At social events, I tend to congregate with my small group of close friends. These are the typical behaviors of a self-proclaimed introvert. Just because you associate yourself with being an introvert doesn't mean you can't turn the tables and be extroverted. I have taught myself to basically imitate extroversion in certain situations when I see opportunities to create mutually beneficial relationships. Now, by saying that I imitate extroversion, I don't mean that I become someone I am not. I am simply saying that I force myself to become slightly uncomfortable, but I never sacrifice my character to find business relationships. So, once you have worked up the courage to encounter other professionals, how do you form relationships that will last?

Be Empathetic

When pursuing a business relationship, I'll often ask people questions about themselves. This is a great tactic because who doesn't like to talk about themselves? It's the one topic that everyone can easily talk about (not to mention, it is often their favorite topic). Don't just ask a deep personal question and then bulldoze the conversation with your own observations—be attentive and listen. Often, we are thinking about our response to their statements, rather than listening, the entire time they are talking. You really need to *listen*, not just hear what the other is saying. This attentiveness will go a long way when forming these relationships in their early stage. Questions like, "Who do you work for?" and "How long have you lived here?" are not terrible questions to start out with, but they will do little to form empathy with the other person. Empathy is the single most important attribute to forming relationships. Simply put, it is being able to put yourself in the others' shoes. Empathy is the only way to form truly meaningful relationships. This is not meant to be a class on social psychology, but I feel like it is necessary to include in this book. Emotional intelligence, or EQ, is important when working your way to the top. If you're still unsure of the whole idea of emotional intelligence, I recommend reading the book *Emotional Intelligence: Why It Can Matter More than IQ* by Daniel Goleman, PhD, which has been cited as being at the forefront of emotional intelligence studies.

Learn about Others

Once you have made the initial contact with someone, start learning more about them. Figure out what they love about their career, what makes them excited to go to work, and what hobbies they enjoy. As I previously mentioned, people enjoy talking about themselves. This should be very easy to do. When you're in their office or place of work, be observant. Notice pictures of family, trophies, or other memorabilia. One technique I've learned to help me keep up with contacts is to write some things about each of them on an index card. I learned this from a mentor of mine at the University of Arkansas. He was one of my favorite professors, an avid businessman, and a passionate leader. The technique goes like this: On each card, write down the person's name and a few facts about them. It doesn't have to be deep and personal; it can be facts like how many kids they have, what their names are, and hobbies they may personally enjoy. Whenever you go to pick up the phone and talk to them, you can ask them about certain hobbies or if their kids are enjoying school. It shows that you care about them on a more personal level. It builds trust and loyalty when you care enough to remember important aspects of their lives. He understood the power of this simple little act. Most people only like to talk about the one thing they know—themselves. When you truly take interest in them, they will think the world of you.

Give More Than You Receive

One rule of thumb I try to keep while forming relationships is to give more than you receive. Give to these contacts without expecting something in return. When you really need their help, they will most likely bend over backward to help you. This is extremely valuable in your career. You never want to be the person who always asks for favors and never gives in return. This will create animosity between you and your contacts. When you give your effort and time generously, the tribal mentality takes over. People will want to go out of their way to help you when you need something because you have shown you're willing to do the same. This brings about a strong sense of camaraderie between you and your contacts. When you begin building these relationships, focus on what you can give them, not what they can bring you. This is the gospel when it comes to strong relationship-building. People often become self-centered when pursuing relationships. They think more about what the other person can do for them than what they can do for the other person.

There are many times when this tactic has rewarded me. Throughout my career, I've established several business contacts that have proven fruitful due to this method. I've helped out one of my longtime clients with several small projects and engineering advice at no cost. These small tasks may take thirty minutes out of my day, but they have paid dividends. He has sent several potential clients my way based on our give-and-take relationship. This small amount of effort and time has produced numerous contracts that were well worth my time. He has even taught me several lessons he's learned over his tenure in real estate development.

Be Active

Over time, your contacts and relationships will slowly drift away if you do not maintain frequency. According to Dr. Jack Schafer, author of *The Like Switch: An Ex-FBI Agent's Guide to Influencing, Attracting, and Winning People Over*, there are four building blocks to creating and maintaining a relationship. The four building blocks are proximity, frequency, duration, and intensity. He declares that if one of these "building blocks" is not maintained, the relationship will suffer or not be created at all. This tip, to be active in creating relationships, is mainly centered on duration and frequency. I strive to contact many of my relationships at least once every couple of weeks. My goal is to contact a majority of my closest contacts at least once a week. A simple phone call is all that is necessary. It is much better to talk to them in person, but a phone call is the next best option. E-mails are a last resort option. With e-mails, you cannot pick up on the visual or verbal nuances that come with face-to-face communication. It is hard to empathize over digital communication because you cannot pick up the nonverbal cues that signal emotions. At the very least, I would recommend e-mailing your contacts at least once every couple of weeks.

Be Authentic

People can spot deception pretty easily. The point of this recommendation is to be yourself. Be honest and up front about your background and goals. Allow yourself to be slightly vulnerable and share personal details that are within good taste. I would never recommend telling your life story, but allow the other person to know where you are coming from and how you became the person you are today.

Now that you know how to form meaningful business relationships, who do you form these relationships with? Like I said before, mutually beneficial relationships, when both parties have something to give, are the most coveted in business. If you are just starting your career, you may be thinking that you don't really have much to bring to the table. This may be true at first, but it is important to focus on what you will bring in the future. I know when I started forming these relationships, I felt somewhat inadequate because my experience was so limited. These relationships should be thought of as more of an investment, though. You must invest time in the relationships for them to grow and become "fruitful."

I have pursued many business relationships with architects, business owners, city staff, bankers, lawyers, and other professionals. These relationships take time to create but can yield huge dividends. These dividends can come in the form of future business, professional support, and reduced interference.

One example of the benefits of business relationships is the associations I create with city staff. As a consulting civil engineer, on any one project I have to answer to many stakeholders—one of them being the city officials. I have done my best to form mutually beneficial relationships with these city employees. I learn everything I can about them, such as how many kids they have, when their birthday is, and what they love about their job. I show genuine interest in their lives. This is not fake; I genuinely do have an interest in how they are doing. If you aren't sincere, others can sense it. I do my best to help them out when I can by giving them the proper documentation in a method that they can process easily. I have volunteered on city boards and genuinely show an interest in their personal and professional lives. By putting this effort forth, I have been paid back graciously by the city staff. They have helped me hit deadlines, alerted me when certain permits are soon to expire, and helped with recommendations to governing bodies such as the city council and planning commission. The relationships were built on mutual benefits, empathy, and genuine curiosity.

Relationship-building is just like any other skill that takes practice and hard work. For some people, it comes more easily than for others. Don't be discouraged if you find it a little tough to begin with. Once you understand the process and the underlying principles stated above, you will have no problems building and maintaining lasting relationships.

CHAPTER 13 SUMMARY

IQ OR EQ? NETWORKING WITH DILBERT

Building relationships can be hard for some engineers, but it doesn't have to be. The reward of creating a social network of peers, mentors, and believers far outweighs the initial discomfort of reaching out. With the following guidelines, you can become a pro at forming these beneficial relationships:

1. Be empathetic.
2. Learn about them.
3. Give more than you receive.
4. Be active.
5. Be authentic.

Whether you're an outgoing socialite or an introverted loner, these tips can be beneficial for creating lasting rapport with the people you meet on your journey to success.

“If a man empties his purse into his head, no one can take it away from him. An investment in knowledge always pays the best interest.”

—BENJAMIN FRANKLIN

CHAPTER 14:

MUCH TO LEARN YOU STILL HAVE . . . THIS IS JUST THE BEGINNING

I REMEMBER IT JUST LIKE it was yesterday. It was a cool Saturday morning on campus. The parking lots were quickly filling up. I could hear the laughter and well-wishes of several families gathering around the arena. The excitement of the event that was about to take place was almost electric. After making plans for after the event with my parents, I walked into the hallway of Barnhill Arena on the beautiful University of Arkansas campus. I tightened my tie and cocked my mortarboard to the side, making sure my tassel was on the correct side. After reminiscing with several classmates about all of the good times we'd had, it was time to shuffle into the crowded event space.

This was the day I was going to get my diploma. This was the culmination of years of hard work, determination, and, of course, the occasional revelry. As I waited to walk across that stage, I thought about all of the work it took to get there. I smiled, thinking I would no longer have silly labs or coursework to complete. Little did I know that walking across that stage was not the end of my educational career but only the beginning. I was truly about to embark on a journey to the real world, where learning must be emphasized daily. If you're not continually growing, you must surely be dying—not necessarily in the physical sense but in the emotional and spiritual sense.

By now, I bet you're wondering if my editors are just terrible when it comes to sentence structure after reading the title of this chapter, but it's no mistake. The title of this chapter comes from a scene in *Star Wars: Episode II - Attack of the Clones.* Yoda is engaged in one of the most intense battles of his life with his nemesis Count Dooku. If you're not a fan of *Star Wars*, this is still one of the greatest fight scenes on film. It's worth a watch on YouTube. At age 874, Yoda still believes there is much to learn. Talk about lifelong learning! I always enjoyed hearing Yoda's wisdom and the backward grammar he used so eloquently. This phrase stood out to me, though. Yoda's pursuit of wisdom enabled him to survive for hundreds of years—900, to be exact. He must have been doing something right. As Yoda would probably express it, "Lifelong learning you must pursue."

As Henry Ford so eloquently stated, "Anyone who stops learning is old, whether at 20 or 80. Anyone who keeps learning stays young." To be able to compete in the workforce, you must never stop pursuing knowledge. This was not a concept I understood straight out of college. I was naïve to think college had taught me everything I would need to know to excel at my job. But why is it important to continue learning? What are the benefits of lifelong learning?

Moore's Law, which proposes that the number of components in integrated circuit chips will increase twofold every eighteen months, has been also attributed to the advancement of technology as a whole. Simply put, technological advances are exponential in nature. Now, this isn't necessarily true for all technological systems, but it's still a metric that many researchers have adopted. In this high-tech age we live in, learning becomes even more important as new systems and technologies constantly emerge. To be a leader in the field, you need to be well versed in the current technology used in your sector. If not, you will get passed over. Technologies become obsolete at almost the same rate as new ones are being created. To give an example, in civil engineering the use of T-squares and light tables was common until very recently. Now all civil engineers use three-dimensional modeling on computers to produce construction plans. One of my favorite quotes on this subject comes from Eric Hoffer, an American philosopher, who stated, "In times of change learners inherit the earth, while the learned find themselves beautifully equipped to deal with a world that no longer exists." This powerful quote underlines the importance of continued education in the fast-changing world we live in.

We, as engineers, understand that it's important to stay on top of technological trends in our field, but why learn about things seemingly unrelated to your field? For engineers, it might seem "beneath" us to study business management, marketing, or public relations, but these are some of the basic skills that help engineers to flourish professionally.

First off, let's define soft skills. According to Dictionary.com, soft skills are "desirable qualities for certain forms of employment that do not depend on acquired knowledge: They include common sense, the ability to deal with people, and a positive flexible attitude." So how do we learn how to utilize these skills when they haven't been adequately discussed for most engineers? My go-to for any idea that I don't thoroughly understand is to read about it. I cannot stress enough how important reading is for personal development! Some great books on the topic of "soft skills" are *Emotional Intelligence* by Daniel Goleman, *The 7 Habits of Highly Successful People* by Stephen Covey, and of course, *How to Win Friends and Influence People* by Dale Carnegie.

> ***"In times of change learners inherit the earth, while the learned find themselves beautifully equipped to deal with a world that no longer exists."***
>
> ***—ERIC HOFFER***

Not only will this skill of lifelong learning benefit your career, it will also benefit your mental health. Lifelong learning will help to keep your mind sharp throughout your life. You can learn obscure skills that bring you happiness. I find that I really enjoy learning new skills. Many of these skills are not practical, but they bring about a sense of accomplishment and gratification. What's a skill you would like to learn? Maybe it's learning how to play the guitar, learning a new language, or even underwater basket-weaving. The benefit does not come from the skill, per se, it comes from the process of learning the skill, not to mention all the interesting topics you could discuss with people based on all the skills you have become proficient in.

Learning can help you live a healthier life as well. By learning about the benefits of good nutrition and exercising, you can implement these into your life to become healthier. This is the same for any habits that are beneficial. I don't think anybody has ever regretted further learning. Now that I believe I have thoroughly explained the reasons to continue learning, let's discuss how.

In the history of human civilization, obtaining knowledge has never been easier. With the invention of the Internet, ideas and products have become readily available all over the world. Never before could you consume knowledge so quickly. You do have to be more selective about the knowledge you receive, but there are many great sources that can be found with the click of a mouse.

The Internet is a great source to learn more from some of the best in the business. It has become a platform for many experts and professionals to dispense their knowledge. Many of the best universities in the world have classes that you can attend online for free! At the time this book was published, you could use the website Coursera.com to take classes from the University of Pennsylvania, Johns Hopkins University, and Stanford University, among many others. Did I mention those classes are free? Even social media sites such as YouTube can be a source of valuable information. I'm not talking about the videos of dogs in crazy outfits; I'm talking about the classes and video logs that many of the greatest entrepreneurial minds upload. You do have to be careful about what knowledge is credible and what you should discard. The Internet is a wild world where anyone can post anything. Unfortunately, some people don't filter this information as well as others. There are many great sources of information online, but that's really just the beginning of your lifelong learning quest.

Let's not forget trade shows and seminars. These are more directed toward your career focus and can be very valuable. At trade shows and seminars, you can get a closer look at the trends within your industry. These seminars are often on the leading edge of technology. How cool would it be to understand the direction your industry is heading? These trade shows give you a first glimpse of technology in beta form. Along with learning at these events comes a great opportunity to network with like-minded professionals. Building up your network at these types of events can give you peace of mind. If something were to happen to your current career or you needed to make a career move, these contacts would become invaluable as a means of obtaining a new place of work.

Along with trade shows come trade journals. Most of these journals are now online, but they still include invaluable information that you can't find just anywhere. Most of these journals come with a subscription cost, but I guarantee that the knowledge you gain from these far outweighs the cost. You can find a trade industry journal on just about anything. Even clowns have a trade journal (believe me, I looked it up). Chances are, you can find a trade journal that will help you see into the future of your industry and allow you to adapt to the ever-shifting trends.

Last, but certainly not least, reading books is a valuable tool for digging deep into the minds of some of the greatest people to ever walk this earth. Where else could you talk with Gandhi, Nelson Mandela, and Arnold Schwarzenegger all in one day (without the use of hallucinogens)? Books are some of the most underrated tools for knowledge you will ever encounter. I'm sure you understand this. If you are reading this nonfiction book, I'm guessing you're not the person I should be preaching to. You're well on your way to great success with this habit. As the famous Stoic philosopher Epictetus once stated, "Books are the training weights of the mind." Continue to read books, and your mind will be constantly strengthened.

As you already know, books are filled with the knowledge of scholars, business moguls, and political leaders. Within books, you can find the lessons these people learned throughout history and how to avoid the pitfalls they succumbed to. Throughout history, the knowledge from these great leaders has been passed down through books. Books are timeless teachers.

So, now you know the importance of lifelong learning. You know what that consists of and how to obtain this knowledge. What if you implemented the philosophy of lifelong learning? For one, you would never be behind in your field no matter how technically oriented it is. Learning keeps you ahead of the curve in all aspects of your career. An inquisitive mind can also open a Pandora's box to all sorts of ideas you didn't think of. Some of these frame shifts can alter your life in ways you didn't think possible.

Through the art of lifelong learning, you can propel yourself light-years ahead of your stagnant peers. You will be able to adapt to change much quicker because you have never stopped learning. By making simple changes in your life, such as adding the habit of going to one seminar a year or reading ten books a year, you can catapult your career ahead of your peers. Out of all the lessons in this book, I think this is the most actionable and beneficial for your life. I challenge you to take a step in learning more in your everyday life, whether that means listening to an audiobook in the car on your way to work (instead of that sports radio talk show with outraged fans) or setting aside fifteen minutes a day to read nonfiction books. These small shifts can produce habits that persist and grow for the rest of your life.

CHAPTER 14 SUMMARY

MUCH TO LEARN YOU STILL HAVE . . . THIS IS JUST THE BEGINNING

So, you thought you were done learning once you walked across that stage at graduation? Think again. If you want to be the best of the best, you must open up to the idea of lifelong learning. In this chapter, the case for learning has been spelled out clearly. I discussed the following items:

1. Why learning long after you earn your degree is extremely important.
2. With the advent of the internet, learning new skills has never been easier.
3. Learning is not limited to reading books, as it can also be done while at conferences and tradeshows.
4. Creating a habit of learning can propel your career lightyears ahead of your colleagues.

You've made an important step by picking up this book with the intent to further your career. However, you cannot let your momentum stall here. Create a daily habit of learning and you will reap the rewards that lifelong learning can produce in your life.

"Moral authority comes from following universal and timeless principles like honesty, integrity, treating people with respect."

—STEPHEN COVEY

CHAPTER 15:

ETHOS IN THE TWENTY-FIRST CENTURY

THE WORD "ETHOS" WAS described in Aristotle's *Rhetoric II* as a person's character. When attempting to persuade others, Aristotle believed that the speaker had to establish ethos, logos, and pathos. Ethos was believed to be the most important of the three appeals. I know you're probably having flashbacks to middle-school English class, but there's no assignment included with this book. Ethos, simply put, is meant to appeal to the audience by showing your credibility. This is done by building trust with your audience and showing your experience in the topic you are presenting. In the twenty-first century, this term has been described as "the distinguishing character, sentiment, moral nature, or guiding beliefs of a person, group, or institution." This use of the word aligns with the idea of ethics in the workplace.

Imagine that one day the company you have worked for your entire working career goes defunct. Your pension plan, and therefore your retirement, is wiped out almost completely in a matter of months. You find out that unethical business practices by upper management are to blame. How are you feeling right now? A little angry? Face it, we'd all be livid if this happened to us. Well, this very thing happened to the many people employed by energy giant Enron Corporation in 2001.

Formed in 1985, Enron was created by the merger of Houston Natural Gas Company and InterNorth Incorporated. With Kenneth Lay as CEO, the company was aiming for greater success. Enron soon became the leader in energy trading and supplying. The early 1990s was a hotbed for growth in business with the introduction of the dot-com era. By the conclusion of the 1990s, the dot-com bubble was in full swing. It was about to pop, and as most Americans were riding high, they were unaware of the catastrophe that was unraveling before them.

Enron was in rapid-growth mode by this time. Constantly increasing its reach, it had started an online trading platform for energy futures in 1999. In each one of the trades performed on this platform, Enron was either the buyer or seller (i.e., the counterparty). This created a false security within its trading partners. With the success of this online trading platform fueling expansion, the company decided to build a high-speed telecommunications network. According to Tom Gros, Enron Communications vice president for global bandwidth trading, "The trading of bandwidth will supercharge the entire internet industry by dramatically increasing the efficiency of bandwidth provisioning and deployment." While his statement was true, this decision was just one more nail in the coffin of the failing company. Having recently been named America's Most Innovative Company by *Fortune* magazine, this was not too surprising of an endeavor. At the time, it seemed like everyone was rooting for Enron to make the jump into telecom. After investing millions in this endeavor, the bottom fell out of the market. The dot-com bubble had finally burst.

American businesses were rocked by this economic downturn. The dot-com bubble is a story of greed and obliviousness. Many investors rode the train of prosperity without fully understanding what train they were on. Many investors threw money at these small start-ups without researching or recognizing their traditional metrics. Enron became very vulnerable, with the majority of its revenue coming from EOL (Enron Online), its online commodities-trading website. This era gave the management of Enron Corporation a decision to make—to either scale down appropriately and lose their benefits or to try to sweep it under the proverbial rug. They chose the latter.

Around late 2000, new CEO Jeffrey Skilling started to explore ways of "hiding" the plight of his enormous company. With the help of his newly promoted CFO, Andrew Fastow, the company effectively swept millions in losses under the rug. They utilized an accounting technique called "market-to-market" accounting. According to Professor Matt Holden of the University of Nevada, Las Vegas, market-to-market accounting "refers to accounting for the value of an asset or liability based on the current market price instead of book value." Put even more simply, the company would create an asset (money-producing investment) and would automatically count its presumed income. If that asset did not produce its projected income, the loss was transferred to another obscure corporation, never showing the loss on their books. Ethical? Definitely not. This smoke-and-mirrors show made it appear to outside investors that the company was far more profitable than the real balance sheets showed.

The accounting firm that handled the Enron account, Arthur Andersen LLP, was also on the hook for this deception. As one of the largest and most trusted accounting firms in the United States, Arthur Andersen was known for superior accounting professionals. Throughout this period of false accounting provided by Enron, Arthur Andersen continued to provide its "stamp of approval" on the Enron financials. As a trusted accounting firm, many investors continued to cling to the approval given by this accounting firm. But soon the veil of deception would be lifted.

CEO Ken Lay jumped ship in the summer of 2001. Months after Skilling took over as CEO, he also left the position for unknown reasons. This became the signal to the outside world that something was amiss. The Securities and Exchange Commission (SEC) soon got involved. According to *Time* magazine in November 2001, "Enron admits accounting errors, inflating income by $586 million since 1997." On December 2, Enron filed for bankruptcy. Soon after its collapse, several of the executives at Enron were under federal indictment, along with the accounting firm Arthur Andersen.

Although you will most likely not wind up in the position of embezzling millions of investor dollars, much can be learned from this story. Whether you're the lowest man on your company's totem pole or the CEO of a major corporation, ethics plays a huge role in how successful you will become.

Let's first delve into the real meaning of this often-misunderstood term. Ethics is defined by Merriam-Webster as "the principles of conduct governing an individual or a group." Put simply, it is a set of often-unwritten rules that individuals follow that govern interactions between two or more parties. This idea is similar to morals but on a more universal scale. Morals are generally attributed to an individual's beliefs, which most often do not translate the same to other individuals. Before I probe too deep into philosophical definitions, I will bring it back to a more generic review. Ethics often take different shapes in different industries, but at the heart of this are a few mutual traits. Every industry has its own code of ethics. In civil engineering, we have the Code of Ethics, produced by the American Society of Civil Engineers (ASCE). The ethics in engineering are often different from business ethics or political ethics, but they all have common threads that we can dissect.

I present to you Matt's Edict of Ethics. I'm still working on the name, but I think it has a nice ring to it. I have compiled a list of ethical traits that I feel are most integral to maintaining order in the marketplace and in society. I have pulled together what I believe to be the most complete and all-encompassing list of behaviors or rules that span all professions. The list has been put together based on my own experiences, as well as the experiences of some of my greatest mentors throughout my career. The list is as follows:

Matt's Edict of Ethics

1. **Conduct affairs with the highest level of integrity**
2. **Respect others**
3. **Maintain a strict standard of excellence**
4. **Hold paramount the safety and well-being of the public**
5. **Further the mission of your chosen profession**

Matt's Edict of Ethics is an abbreviated list of ethics that I believe span the gamut of professions. The code of ethics for each profession will often have similar canons, so I combined certain aspects to create all-encompassing principles. If you maintain these ethics within your career, I have no doubt you will be successful in your endeavors, as well as respected by your peers. In the following paragraphs, the Edict of Ethics will be dissected.

Conduct affairs with the highest level of integrity

Typically, the first canon or principle in a code of ethics is based on integrity or honesty. These terms, often used interchangeably, are integral in fostering a successful market. If there's no trust in the marketplace, transactions will not be executed. Put another way, would you purchase a product or service without expecting the item to perform as presented? For transactions to occur, integrity is key. Unfortunately, business isn't always conducted honestly. As seen in the example above, the executives at Enron did not perform business affairs with integrity and honesty. This unfolded on a large scale, but it can happen on a much smaller scale as well.

In my industry of construction and development, a lack of integrity can, and has many times, resulted in costly disasters. These disasters can result in monetary loss or, even worse, the injury or death of an innocent member of the public. Integrity is paramount in engineering due to the potential impacts to the safety of the public. We will delve further into the impacts on the public with the fourth canon.

Respect others

This principle should be familiar from your kindergarten classroom. The Golden Rule is an understood law that holds the very fabric of society together. The Golden Rule states, "Do unto others as you would have them do unto you." This law of reciprocity creates a fair marketplace for all to do business. The underlying principle to the golden rule is to respect others. Respecting others is not just a good practice; it is paramount when building relationships.

Maintain a strict standard of excellence

If you're reading this book right now, I can imagine this canon is first and foremost for you. The act of reading and learning is one way of maintaining a strict habit of excellence. If we break this principle down to its core, it states that we work hard every day to do the best that we can and nothing less.

> ***"If you can be honest and responsive, then you are already ahead of the game."***
>
> ***—ANTHONY FASANO***

Hold paramount the safety and well-being of the public

The following principle is often not as regarded in business as it is in professions such as engineering, nursing, and law enforcement, but it's still relevant for business professionals. The safety and well-being of the public should be upheld regardless of your profession. Many decisions made in a business can affect the well-being of the public.

In a tragic case of not upholding this principle, hundreds of people were injured and killed when disaster struck the great Hyatt Regency Kansas City hotel. Throughout our engineering classes, we discussed the importance of our decisions in designing buildings, roadways, bridges and other items. The case of negligence that was most commonly discussed was the disaster that occurred at the Hyatt Regency in Kansas City.

On July 17, 1981, a structural connection located within a revolutionary skywalk in the hotel failed. This connection held the second- and fourth-floor skywalks suspended above a giant atrium lobby. During a large social event at the hotel, many of the guests were enjoying their evening when the second- and fourth-floor skywalks came crashing down under the weight of the patrons reveling on them. The resulting catastrophe took the lives of over one hundred visitors and injured more than two hundred. How did this happen? What caused such a horrific event to unfold?

Though blame for this event was thrown back and forth, the engineer of record did have his professional stamp on the plans for this project and was ultimately responsible for the design. Although he may not have designed the actual connections that failed, he was ultimately responsible for the safety and well-being of the public. While the designs seem fairly similar, this slight difference was all that was needed for failure. I rarely deal in structures, but this is still a distressing story to hear. You often forget that anything you do makes an impact on others' lives. On paper, your design, tax filing, or executive order may seem harmless. But when brought to life, will it adversely affect people? To what degree? These are all questions you should ask yourself before you make important decisions that could possibly affect the health and well-being of the public.

Further the mission of your chosen profession

Engineers are very proud of their profession. It takes countless hours of hard work to call yourself an engineer. This can also be said of many of the other professions out there. Your goal should be to further the mission of your profession in everything you do. The more you establish the credibility of your profession, the more people will respect your profession.

There may be many times in your career when you feel pressure to cut corners or take unethical shortcuts. These pressures could come from budget cuts, external burdens, or even upper management. To maintain your integrity, you must resist these temptations. Ultimately, you will be respected for your willpower, even though it may be tough in the moment.

These are no easy tasks. Often, the consequences of a rash or immoral decision are hard to see in the moment. Once you have created a set of values for yourself, you will find this much easier. When you have a clear set of values, you can easily back-check your decision against your values and ethics to see if it passes the test. This will make your decision-making process much more streamlined. The world of business makes ethical decision-making tough due to the financial consequences of such decisions, but with a clear understanding of your ethical and moral principles, you can make hard decisions much easier.

CHAPTER 15 SUMMARY

ETHOS IN THE TWENTY-FIRST CENTURY

Nearly all of the decisions you make in your career touch others. To ensure that your decisions don't affect others negatively, you must uphold ethical standards for yourself. If you need help identifying ethical behavior, give Matt's Edict of Ethics a try!

Matt's Edict of Ethics

1. Conduct affairs with the highest level of integrity
2. Respect others
3. Maintain a strict standard of excellence
4. Hold paramount the safety and well-being of the public
5. Further the mission of your chosen profession

When you struggle with a decision, ask yourself first if you would feel guilty describing the choice to your children. If you don't have kids, imagine your mother seeing the effects of your decision in the newspaper. This usually puts your decision in perspective and often makes it easier to make the right one.

"I shall be telling this with a sigh

Somewhere ages and ages hence:

Two roads diverged in a wood, and I—

I took the one less traveled by,

And that has made all the difference."

—ROBERT FROST

CONCLUSION

THROUGH THE PROCESS OF research and discovery in writing this book, I have learned much about the habits and lifestyles of some of the greatest engineers-turned-businesspeople. Although they all had very different backgrounds, they exhibited many of the same traits laid out in this book. Their fearless personalities helped to shape the course of history as we know it.

Because when you get down to it, that's what it's all about—becoming fearless or uncomfortable in the pursuit of something you believe in. In many ways, this book is the fearless mission I am embarking on. It has been a truly eye-opening experience. I have opened myself up to the world, which is pretty uncomfortable, to say the least. You can exhibit all of the traits extolled in this book and still not be successful if you can't go out on a limb in the hunt for success in your dreams.

You have completed the journey throughout this book with me, and I commend you for your persistence. You have shown you can beat the odds. While a majority of people don't finish a book they started, you have risen above the masses. I hope you have enjoyed this book so much you have dog-eared pages that were important to you and written pertinent messages to yourself all through their once-clean margins. I hope you keep coming back to this book for advice as it sits on your bookshelf or in your Kindle, such as you would with an old friend you turn to for guidance.

My wish for you after reading this book is a deeper appreciation for the skills necessary to be successful, not only in business but also in life. You have now gotten a chance to sit among the greats who have made this world a better place in engineering, as well as business. My hope is that you take the skills discussed in this book and build upon them. Further your education and pursue your greatest potential. My wish is that you take a leap of faith and follow your passions, because nobody who ever abandoned their dreams made it into the history books.

Now that you know what the Business of Engineering is really about, you can see that it can be employed in any profession. In this age of globalization and international business, we are all in the business of something. You cannot avoid business in this service market. We are all being asked to adapt to the new frontier and gain the skills necessary to compete on all fronts. My guarantee for you is that if you take the path of greater education, you will not regret your investment. To compete in this society, learning and implementing new skills will put you at the front of the pack (even if that pack may include robots with artificial intelligence).

I appreciate you taking time out of your often-busy life and learning alongside me as we explored deep into the skills of business success. It has truly been an honor for me to have you read this book, and I would like to thank you personally for letting me walk with you on your journey to success! Before we close, I would just like to leave you with one more quote to think about before you shut this book:

> ***"Successful people do what unsuccessful people are not willing to do. Don't wish it were easier; wish you were better."***
>
> **—JIM ROHN**

REFERENCES

Chapter 1: What Is the Business of Engineering?

"Job Outlook 2016: The Attributes Employers Want to See on New College Graduates' Resumes." http://www.naceweb.org/career-development/trends-and-predictions/job-outlook-2016-attributes-employers-want-to-see-on-new-college-graduates-resumes.

Berger, Guy. "Data Reveals the Most In-demand Soft Skills Among Candidates." LinkedIn Talent Blog. LinkedIn. 30 Aug. 2016. http://business.linkedin.com/talent-solutions/blog/trends-and-research/2016/most-indemand-soft-skills.

"Thomas Edison." *Biography.com*. A&E Networks Television. 30 Nov. 2016. Web. 16 Feb. 2017. http://www.biography.com/people/thomas-edison-9284349#synopsis.

"Thomas Edison." *Biography.com*. A&E Networks Television. 30 Nov. 2016. Web. 16 Feb. 2017. http://www.biography.com/people/thomas-edison-9284349#synopsis.

"Henry Ford." *Encyclopædia Britannica*. Encyclopædia Britannica Inc., n.d. Web. 16 Feb. 2017. https://www.britannica.com/biography/Henry-Ford.

"Elon Musk." *Forbes*. n.d. Web. 16 Feb. 2017. http://www.forbes.com/profile/elon-musk.

"Elon Musk." *Biography.com*. A&E Networks Television. 21 Nov. 2016. Web. 16 Feb. 2017. http://www.biography.com/people/elon-musk-20837159#early-life.

Gregersen, Erik. "Elon Musk." *Encyclopædia Britannica*. Encyclopædia Britannica Inc. n.d. Web. 16 Feb. 2017. https://www.britannica.com/biography/Elon-Musk#ref1121137.

Rainie, Lee and Andrew Perrin. "Slightly Fewer Americans Are Reading Print Books, New Survey Finds." *Pew Research Center*. N.p., 19 Oct. 2015. Web. 16 Feb. 2017. http://www.pewresearch.org/fact-tank/2015/10/19/slightly-fewer-americans-are-reading-print-books-new-survey-finds.

Bender, Joshua. "Topic: Amazon." *www.statista.com*. N.p., 05 Jan. 2017. Web. 16 Feb. 2017. https://www.statista.com/topics/846/amazon.

"Jeff Bezos." *Biography.com*. A&E Networks Television. 18 July 2016. Web. 16 Feb. 2017. http://www.biography.com/people/jeff-bezos-9542209#synopsis.

"Jeff Bezos." *Entrepreneur*. N.p., 10 Oct. 2008. Web. 16 Feb. 2017. https://www.entrepreneur.com/article/197608.

Rafferty, John P. "Mary Barra." *Encyclopædia Britannica*. Encyclopædia Britannica Inc. 06 Mar. 2015. Web. 16 Feb. 2017. https://www.britannica.com/biography/Mary-Barra.

"Moment Distribution Method for Continuous Beams." *Moment Distribution Method for Continuous Beams—EngineeringWiki*. N.p., n.d. Web. 16 Feb. 2017. http://www.engineeringwiki.org/wiki/Moment_Distribution_Method_for_Continuous_Beams.

Chapter 2: The Engineer of the Future

"Read 'The Engineer of 2020: Visions of Engineering in the New Century' at NAP.edu." National Academies Press: OpenBook. The National Academies Press. 2004. http://www.nap.edu/read/10999/chapter/6.

Smith, Ralph J. "Engineering." *Encyclopædia Britannica*. Encyclopædia Britannica Inc. 5 Oct. 2017. http://www.britannica.com/technology/engineering.

Chapter 3: Mercedes-Benz and the Path to Success

Maranzani, Barbara. "Bertha Benz Hits the Road." *History.com*. A&E Television Networks. 5 Aug. 2013. http://www.history.com/news/bertha-benz-hits-the-road.

The Editors of Encyclopædia Britannica. "Karl Benz." *Encyclopædia Britannica*. Encyclopædia Britannica Inc. 15 Jan. 2015. http://www.britannica.com/biography/Karl-Benz.

Huber, Martin Fritz. "A Brief History of the Sub-4-minute Mile." Outside Online. 9 June 2017. http://www.outsideonline.com/2191776/brief-history-sub-4-minute-mile.

"First Four-minute Mile." *History.com*. A&E Television Networks. 2010. http://www.history.com/this-day-in-history/first-four-minute-mile.

"Environment." *Merriam-Webster.com*. Merriam-Webster. n.d. Web. Jan. 26, 2018.

Chapter 6: Dopamine and the Structural Beam Problem

Michigan State University. "Brief Interruptions Spawn Errors." *MSUToday*. N.p., n.d. Web. 16 Feb. 2017. http://msutoday.msu.edu/news/2013/brief-interruptions-spawn-errors.

Zamora, Dulce. "Sleep Deprivation at the Workplace." *WebMD*. WebMD. n.d. Web. 16 Feb. 2017. http://www.webmd.com/sleep-disorders/features/sleep-deprivation-workplace#1.

Chapter 7: Ingenuity and the Engineer-How to Harness Creativity

Wiseman, Liz. *Rookie Smarts: Why Learning Beats Knowing in the New Game of Work*. New York: Harper Business. 2014. Print.

Chapter 8: The Impact of the "Zero Hour" and Utilizing Crunch Time

"Andersonville Prison." *Civil War Trust*. Civil War Trust. n.d. Web. 16 Feb. 2017. http://www.civilwar.org/education/history/warfare-and-logistics/warfare/andersonville.html.

University of Chicago Press Journals. "Getting Things Done: How Does Changing the Way You Think About Deadlines Help You Reach Your Goals?" ScienceDaily. 26 Aug. 2014. http://www.sciencedaily.com/releases/2014/08/140826121054.htm.

"Plan." *Merriam-Webster.com*. Merriam-Webster. n.d. Web. Feb. 23, 2017.

Chapter 9: Blood, Sweat, and Engineers

Conley, Mikaela. "Persistence Is Learned from Fathers, Says Study." *ABC News*. ABC News Network. 15 June 2012. Web. 02 Mar. 2017. http://abcnews.go.com/Health/persistence-learned-fathers-study/story?id=16571927.

Kim, Larry. "30 Inspiring Billion-Dollar Startup Company Mission Statements." *Inc.com*. Inc. 05 Nov. 2015. Web. 03 Mar. 2017. http://www.inc.com/larry-kim/30-inspiring-billion-dollar-startup-company-mission-statements.html.

Wing, Rena R. and Robert W. Jeffery. "Benefits of Recruiting Participants with Friends and Increasing Social Support for Weight Loss and Maintenance." *Journal of Consulting and Clinical Psychology* 67.1 (1999): 132–38. Web. 3 Mar. 2017.

TED Talks Director. YouTube. 9 May 2013, http://www.youtube.com/watch?v=H14bBuluwB8. Accessed 8 Sept. 2017.

Chapter 10: Sun Tzu and Commanding the Apollo 13: The Art of Pure Leadership

Dunbar, Brian. "Apollo 13." *NASA*. NASA. 06 June 2013. Web. 16 Feb. 2017. https://www.nasa.gov/mission_pages/apollo/missions/apollo13.html.

NASA. NASA. n.d. Web. 16 Feb. 2017. https://www.hq.nasa.gov/alsj/a13/a13_LIOH_Adapter.html.

Tzu, Sun. *The Art of War*. Lexington, KY: Filiquarian. 2015. Print.

Chapter 11: Thou Shalt Not Simply Trot Out Thy Usual Shtick

Shandrow, Kim Lachance. "7 Powerful Public Speaking Tips From One of the Most-watched TED Talks Speakers." *Entrepreneur*. N.p., n.d. Web. 17 Feb. 2017. https://www.entrepreneur.com/article/239308.

Feloni, Richard. "Tony Robbins Shares His 3 Best Public Speaking Tips." Business Insider. Business Insider. 21 Nov. 2014. Web. 17 Feb. 2017. http://www.businessinsider.com/tony-robbins-public-speaking-tips-2014-11.

North, Marjorie. "10 Tips for Improving Your Public Speaking Skills." Harvard Professional Development. Harvard DCE." *Forbes*. N.p. 03 Nov. 2016. Web. 17 Feb. 2017. http://www.extension.harvard.edu/professional-development/blog/10-tips-improving-your-public-speaking-skills.

Carnegie, Dale and J. Berg. Esenwein. *The Art of Public Speaking*. Place of Distribution Unknown: Editoria Griffo, 2015. Print.

Chapter 12: The Nike Swoosh and Marketing Your Own Brand

"Our History in Depth—Company—Google." *Our History in Depth—Company—Google*. N.p., n.d. Web. 16 Feb. 2017. https://www.google.com/about/company/history.

Levinson, Philip. "How Nike Almost Ended Up with a Very Different Name." *Business Insider*. Business Insider, 19 Jan. 2016. Web. 20 Feb. 2017. http://www.businessinsider.com/how-nike-got-its-name-2016-1.

Chapter 13: IQ or EQ? Networking with Dilbert

"Empathy." *Merriam-Webster.com*. Merriam-Webster. n.d. Web. 31 Dec. 2016.

Chapter 14: Much to Learn You Still Have . . . This Is Just the Beginning

Chandler, David L.. "How to Predict the Progress of Technology." *MIT News*. N.p. 06 Mar. 2013. Web. 16 Feb. 2017. http://news.mit.edu/2013/how-to-predict-the-progress-of-technology-0306.

"Soft Skills." *Dictionary.com*. Dictionary.com. n.d. Web. 16 Feb. 2017. http://www.dictionary.com/browse/soft-skills.

Chapter 15: Ethos in the Twenty-First Century

"Enron to Trade Bandwidth." *CNNMoney*. Cable News Network, n.d. Web. 16 Feb. 2017. http://money.cnn.com/1999/05/20/technology/enron.

Staff, Investopedia. "Mark To Market—MTM." *Investopedia*. N.p. 23 Jan. 2014. Web. 16 Feb. 2017. http://www.investopedia.com/terms/m/marktomarket.asp.

"Definition of 'Mark to Market Accounting'—NASDAQ Financial Glossary." *NASDAQ.com*. N.p., n.d. Web. 16 Feb. 2017. http://www.nasdaq.com/investing/glossary/m/mark-to-market-accounting.

"Ethic." *Merriam-Webster.com*. Merriam-Webster, n.d. Web. 28 Jan. 2017.

Babar, Aditya. "Hyatt Regency Walkway Collapse: Did the Structural Analysis Go Wrong?" *Indovance.com*. N.p., 05 Nov. 2016. Web. 16 Feb. 2017. https://www.indovance.com/knowledge-center/hyatt-regency-walkway-collapse-did-the-structural-analysis-go-wrong.

"Ethos." *Merriam-Webster.com*. Merriam-Webster. n.d. Web. 11 Feb. 2017.

ABOUT THE AUTHOR

MATTHEW K. LOOS, P.E. graduated from the University of Arkansas with a Bachelor of Science in Civil Engineering with a minor in General Business. He has worked in the civil engineering consulting field since 2012. He specializes in commercial real estate development throughout the state of Texas. His projects include, but are not limited to multi-family residential, retail, campus, and office developments. As a business minor at the Sam M. Walton College of Business at the U of A, Loos was given a unique taste of the vast landscape of business. This interest in business strategies has only increased since his graduation and was a driving force behind the writing of this book.

Made in the USA
Lexington, KY
08 November 2019